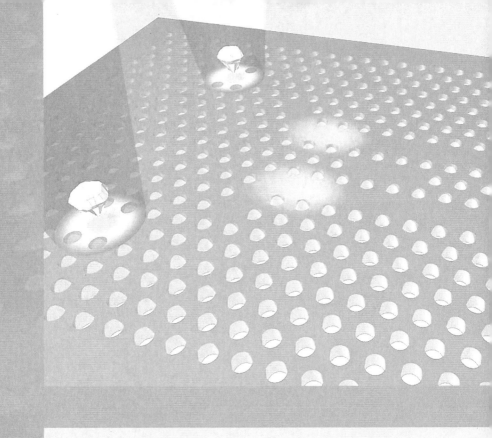

Integrated Quantum Hybrid Systems

Integrated Quantum Hybrid Systems

Janik Wolters

PAN STANFORD PUBLISHING

Published by

Pan Stanford Publishing Pte. Ltd.
Penthouse Level, Suntec Tower 3
8 Temasek Boulevard
Singapore 038988

Email: editorial@panstanford.com
Web: www.panstanford.com

British Library Cataloguing-in-Publication Data
A catalogue record for this book is available from the British Library.

Integrated Quantum Hybrid Systems

ISBN 978-981-4463-82-9 (Hardcover)
ISBN 978-981-4463-83-6 (eBook)

Printed in the USA

Contents

PART II: QUANTUM SYSTEMS FOR INTEGRATION INTO HYBRID DEVICES

PART IV: COUPLING OF QUANTUM SYSTEM TO OPTICAL MICROSTRUCTURES

Foreword

Max Planck's ingenious introduction of the quantum of action h marked the start of quantum theory more than hundred years ago. Research in the first half of the twentieth century was then driven basically by developing the full framework and by suggesting an interpretation of quantum theory. The second half witnessed an ever-increasing quest to transfer the laws of quantum physics into a novel technology. However, only within the last few decades has significant progress been made. The advancements of laser technology and microscopy brought us light sources and detectors that enabled us to optically probe single-quantum systems. In addition, the progress in the growth of ultra-pure optical material and its nanostructuring in modern clean-room facilities have relaxed the need to keep quantum systems in exotic environments in order to maintain their peculiar properties.

Today, studying and manipulating single-quantum systems in the lab is routine. It is even possible to assemble first optical quantum elements consisting of a few quantum systems interacting via exchange of individual photons.

The goal of integrated quantum technology is now to integrate more and more emitters from various materials and different miniaturized photonic structures on a controllable hybrid platform. Such optical quantum devices will help in the future to speed up and secure information processing or to perform sensing tasks with a precision far beyond classical limits.

In his book, Janik Wolters gives a comprehensive overview of the field of integrated quantum hybrid systems. Starting from the very fundamentals of quantum optics, we learn how individual building blocks of a possible quantum technology platform can be isolated, studied, and understood in detail. Moreover, the author introduces the unique optical properties of micro- and nanostructures and explains how they can be fabricated with structure sizes of only tens of the wavelength.

Out of the variety of presently pursued approaches, the author focuses on color defect centers in diamond. This specific system provides the outstanding advantage of optical stability and photon emission even at room temperature. In addition, defect centers can host individual electron spins that are easily initiated, manipulated, and read out optically. By this, an intrinsically quantum feature— the spin—can be integrated into the functionality of an integrated quantum element.

Janik Wolters successfully combines introductions to both quantum optics and nanotechnology, which is clearly essential to understand hybrid quantum systems. In a superb way, he complements the introductory parts which are accessible to beginners in the field with brand-new and exciting results fresh from the lab. Examples are the first demonstration of the quantum Zeno dynamics observed on a single spin as well as Purcell enhancement of single photon emitters in a photonic crystal resonator.

The book not only provides an interesting update to experts in the field, but also motivates students to really get into the exciting field of integrated quantum hybrid systems.

Oliver Benson
Jelena Vuckovic
Berlin and Stanford, CA
February 2015

Preface

Integrated quantum hybrid devices, built from classical dielectric nanostructures and individual quantum systems, promise to provide a scalable platform to study and exploit the laws of quantum physics. On the one hand, there are novel applications, such as efficient computation, secure communication, and measurements with unreached accuracy. On the other hand, hybrid devices might serve to explore the limits of our understanding of the physical world, i.e., the formalism of quantum mechanics. Thus, optical quantum hybrid systems have got into the focus of many researchers worldwide.

The present work gives a comprehensive introduction into this exciting and fast growing field. Several novel experimental results are presented and and new proposals for further studies are discussed.

Starting with the quantization of the electromagnetic field, a solid theoretical basis to understand light–matter interaction is elaborated in Part I. In Part II follows a description of several solid-state quantum systems, namely quantum dots, organic molecules, and color centers in diamond, which are all possible constituents of integrated hybrid devices. Particular attention is paid to experiments with the negatively charged nitrogen-vacancy center in diamond, where many achievements were reached in the recent years. Some examples are spectral diffusion measurements, as well as coherent spin manipulation and its inhibition by the quantum Zeno effect.

Part III introduces dielectric micro- and nanostructures as second ingredient for building integrated hybrid devices. Here, the focus is on photonic crystals, of which not only theoretical and experimental fundamentals, but also a few classical applications, like refractive index measurement and thermo-optical switching, are discussed.

Finally, the actual quantum hybrid systems are introduced in Part IV. Here, fundamental methods and recent experimental results are discussed. For example, novel experimental results on the controlled coupling of single nitrogen-vacancy centers to phonic crystal cavities are presented. These results pave the way for more complex devices and can help realize schemes to entangle distant nitrogen-vacancy centers on-chip. Once these schemes are successfully implemented, they will serve to fully exploit the promises and prospects of an integrated quantum technology platform.

In the first place it is fair to state that we are not experimenting with single particles, any more than we can raise Ichthyosauria in the zoo.

—E. Schrödinger, 1952 [1]

Über Halbleiter sollte man nicht arbeiten, das ist eine Schweinerei, wer weiss, ob es überhaupt Halbleiter gibt.

One should not work on semiconductors, that's a mess, who knows whether there are semiconductors at all.

—W. Pauli, 1931

Chapter 1

Introduction

The two preceding quotations of Schrödinger and Pauli from the first half of the 20th century might suggest the opposite, but it is in fact research on individual quantum objects and on semiconductors which has seen tremendous progress in the last decades. Single quantum systems are not anymore only a theoretical concept that is applied in Gedanken experiments, but single atoms, ions, photons, or individual solid state quantum systems are frequently used in many real-world experiments.

Among the quantum systems, photons as quanta of the electromagnetic field have a particular role. On the one hand they represent interesting quantum particles themselves that are used in many proposals and experiments on quantum communication and computation. On the other hand they are often the only messengers, carrying information about their emitter. For example, the development of the quantum theory of the hydrogen atom, led by Bohr, Heisenberg, Schrödinger, Sommerfeld, and Dirac in the early 20th century, was driven by precise measurements of the emission wavelength by Lorentz, Zeeman, Stark, Lamb, and others (cf. Table 1.1). In the early days these observations required large ensembles of atoms or ions in gas cells. Later on, the number of required ions could be strikingly reduced with the invention of the

Integrated Quantum Hybrid Systems
Janik Wolters
Copyright © 2015 Pan Stanford Publishing Pte. Ltd.
ISBN 978-981-4463-82-9 (Hardcover), 978-981-4463-83-6 (eBook)
www.panstanford.com

Table 1.1   List of Nobel laureates relevant to the field

Year	Name(s)
1902	H. A. Lorentz, P. Zeeman
1918	M. K. E. L. Planck
1919	J. Stark
1921	A. Einstein
1922	N.H. D. Bohr
1929	P. L.-V. P. R. de Broglie
1932	W. K. Heisenberg
1933	E. Schrödinger, P. A. M. Dirac
1944	I. I. Rabi
1945	W. Pauli
1954	M. Born
1955	W. E. Lamb
1956	W. B. Schokley, J. Bardeen, W. H. Brattain
1964	C. H. Townes, N. G. Basov, A. M. Prokhorov
1966	A. Kastler
1981	N. Bloembergen, A. L. Schalow
1989	N. F. Ramsey, W. Paul
1997	S. Chu, C. Cohen-Tannoudji, W. D. Phillips
2000	Z. I. Alferov, H. Kroemer, J. S. Kilby
2005	R. J. Glauber, J. L. Hall, T. W. Hänsch
2009	C. K. Kao
2012	S. Haroche, D. J. Wineland

According to the committee's website nobelprize.org, 2014 W. Moerner, S. W. Hell at the time of writing.

Paul trap in the 1950s [2]. Since then, ions in the gas phase played a pioneering role in quantum optics: already by the end of the 1980s they could be observed on the single emitter level [3]. This was rapidly followed by the observation of single organic molecules [4], and in the early 2000s by single quantum dots [5], defects in diamond [6], and neutral atoms [7].

The tremendous progress was founded on two enabling technologies. One is the laser, which was developed since its invention by Maiman in 1960. With so far unreached spectral brightness it allows to selectively address narrow transition lines with high intensity, without generating too much background light. Furthermore, on the detection side, fast and efficient single-photon detectors, namely photomultiplier tubes and avalanche photodiodes, became widely available. This happened mainly due to the enormous progress through which material science and semiconductor electronics went.

Starting with the development of the first bipolar junction transistor by Shockley, Bardeen, and Brattain in 1948 [8], the first integrated circuit was built by Kilby and Noyce in 1958. Today, very large scale integration (VLSI) has reached a level of billions of transistors monolithically integrated on a single chip. For further integration, the energy-efficient interconnection of different units is a major hurdle. As a solution, optical links are promising. Such optical links are already widely used in telecommunication and are based on two landmarking technologies: semiconductor lasers, which were first put forward by Alferov and Kroemer, and optical fibers, initially investigated by Kao [9]. For on-chip integration, these optical interconnects demand not only highly miniaturized light sources but also wave guiding and routing. Several concepts have been proposed and approved in the recent years to address the latter task. Here index waveguides, being closely related to the optical fiber, and photonic crystals, in which light is confined and guided due to photonic band gaps, are notable. For efficient integrated light sources, existent devices must be miniaturized. At the ultimate level this ends up with single emitters on-a-chip, and allows for a device operating with single quanta.

The energy efficiency of single quantum devices would be enormous, and potentially the laws of quantum mechanics allow further advantages: single photons can be used for unconditional secure communication, quantum computing might allow for exponential speed-up in solving complex problems, and sensors can reach extreme accuracy and sensitivity. For these future applications it is not sufficient to understand and handle the individual constituents, i.e., light emitters and light-guiding structures, but rather their intricate interaction. In addition, the technology for an integrated quantum photonic platform must be developed and controlled. Furthermore, strong light–matter interaction in nanostructures gives rise to manifold effects and rich physics in integrated quantum hybrid system.

To exploit all these prospects of an integrated quantum technology platform, a profound understanding of the hybrid system is indispensable. Here research on large multipartite quantum systems is still at the beginning, in particular regarding large entangled systems. So far, well-isolated atoms and ions in the gas phase again have played a pioneering role, and entanglement of up to 14 ions could be demonstrated [10]. Nevertheless, because of the required complicated trapping and cooling mechanisms, it is questionable whether the gas-phase experiments can be advanced to build a scalable platform or field applications. In contrast, solid state systems avoid these problems and promise scalability and robustness at the same time.

This work gives a comprehensive introduction to the growing field of integrated quantum hybrid systems. Novel experimental results are presented, and new proposals for further studies are discussed. The text is organized in four parts, which are briefly summarized below.

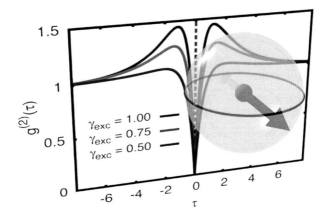

Figure 1.1 Illustration of anti-bunched photon emission from a three-level system and the Bloch sphere representing a spin-1/2 system. Both are key results from Part I.

Part I is devoted to the *theory* of quantum electrodynamics. The fundamental properties of the quantized electromagnetic field are discussed at the beginning, before light–matter interaction is treated. The latter starts with a discussion of the dynamics of quantum systems using second-order perturbation theory, i.e., Fermi's golden rule. Here, fundamental phenomena like single-photon emission are predicted. Then, the semiclassical theory of light–matter interaction is developed, leading to the Bloch equations and Rabi oscillations. At the end of part I the fully quantized Jaynes–Cummings model of cavity quantum electrodynamics is introduced. One key result from this is the Purcell effect, which is found in the limit of strong cavity damping.

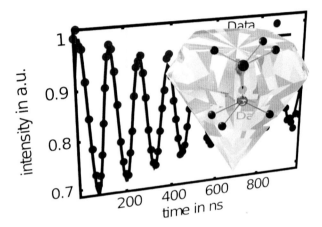

Figure 1.2 Artist's view of a nitrogen-vacancy center in diamond, which hosts a single electron spin. This spin is used to experimentally demonstrate Rabi oscillations and to study decoherence of the dynamics of a single quantum system, a key result in Part II.

Part II introduces common optically active solid state quantum systems, which are suitable for integration into quantum hybrid devices. Quantum dots and organic molecules are treated only briefly, while the focus of this section is on defects in diamond. In particular the nitrogen-vacancy center in diamond is regarded. This center is used to *experimentally* investigate many of the effects theoretically predicted in Part I. Its electron spin is used to demonstrate coherent spin manipulation and its inhibition via the quantum Zeno effect, while the optical transition is used for single-photon generation. In the latter, the focus is on the spectral stability of the emission, which turns out to be a major hurdle for integrated quantum optics experiments.

Part III subsequently treats optical microstructures. Starting with classical electrodynamics, the fundamental mechanisms of total internal reflection, index guiding, and photonic bandgaps are introduced. Later on, these are applied to achieve light guiding and confinement. Here, experimental results on index guiding structures like disk resonators and strip waveguides are presented, as well as results on photonic bandgap structures, the so-called photonic crystal cavities. The latter ones are also used to demonstrate applications like thermo-optical switching or refractive index measurements.

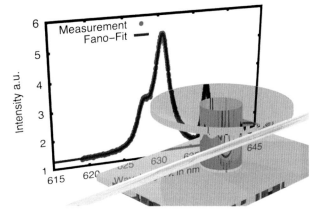

Figure 1.3   Dielectric microstructures. Illustration of a microdisk resonator coupled to a waveguide as one example of the systems studied in Part III. The measured spectrum shows pronounced resonances of a photonic crystal cavity.

Part IV treats the coupling of quantum systems to optical microstructures. Here the technological approaches to build quantum hybrid systems and experimental key results like Purcell enhancement of single nitrogen-vacancy centers and strong coupling are introduced. Later on, two proposals for cavity-enhanced on-chip entanglement are discussed. These schemes, which are scalable to several emitters, illustrate the prospects of an integrated quantum hybrid platform.

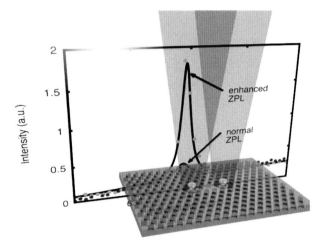

Figure 1.4 Prospects of quantum hybrid systems as discussed in Part IV. Experimentally observed Purcell enhancement of a single nitrogen-vacancy center coupled to a photonic crystal cavity and coupling of two nitrogen-vacancy centers via a shared cavity mode.

The work is completed by a concluding chapter and an appendix with a brief summary of the author's contributions to the field, an extensive bibliography, lists of figures, tables and abbreviations, and an index.

PART I

FUNDAMENTALS OF QUANTUM OPTICS

Introduction

In Part I, several fundamental results of theoretical quantum optics and light–matter interaction are elaborated. It provides all necessary theoretical background to understand the physics of the single quantum emitters introduced in Part II and later on coupled systems discussed in Part IV.

To provide a solid basis, the normal modes of the electromagnetic field and the classical minimal coupling Hamiltonian are introduced in Chapter 2. From the classical Hamiltonian the Newton–Lorenz equation and Maxwell's equations are derived using the Hamilton formalism. The latter ones are particular important for Part III, where dielectric microstructures are treated. After this instructive exercise, the canonical quantization is applied to obtain a quantized formulation with Hamilton operator and canonical commutator relations. Subsequently, the decomposition of the Hamiltonian into the particle, radiation, and light–matter interaction parts is motivated.

It is assumed that the solution of the particle Hamiltonian are well known from basic quantum mechanics textbooks and do not need further discussion. Chapter 3 leads directly to a revision of the basic properties of the unperturbed quantized electromagnetic field. Here, the operators corresponding to the classical fields are introduced, before the properties of Fock states and coherent states are briefly examined.

Using the background from the previous chapters, light–mater interaction is discussed in Chapter 4. Starting with Fermi's golden rule, the emission and absorption mechanisms are studied. Particularly, the latter one leads to a discussion of photon detection mechanisms and photon statistics. Furthermore, Fermi's golden rule motivates a rate equation model of two- and three-level systems, which are discussed extensively, yielding to an understanding of the fundamental physics of individual quantum emitters.

Later on, coherent interactions are considered in the semiclassical approach, resulting in the optical Bloch equations and their solution, i.e., the well-known Rabi oscillations for two-level systems. This is extended to the interesting case of three-level systems in a Λ configuration before the fully quantized Jaynes–Cummings model is introduced at the end of Part I. Here, the principle results of cavity quantum electrodynamics, i.e., vacuum Rabi splitting and oscillations, as well as the Purcell effect are derived.

Chapter 2

From Classical Electrodynamics to the Quantized Hamiltonian

In this chapter the dynamics and properties of the electromagnetic field and charged particles are elaborated. For this, charges and normal modes, as well as their fundamental properties are introduced in Section 2.1, followed by a derivation of the classical equations of motion in the framework of the Hamilton formalism in Section 2.2. Maxwell's equations are one important result from this section. At the end of this chapter, in Section 2.3, the Hamiltonian is quantized, giving the basis for a further discussion of quantum electrodynamics phenomena.

2.1 Charged Particles and Normal Modes

Classical non-relativistic electrodynamics describes experiments, where *electrically charged particles* interact with each other and the complex *normal modes* of the electromagnetic field.

The particles have a spatial extension much smaller than all other considered length scales and thus can be considered to be point-like. Each particle μ carries the scalar *electric charge* q_μ, the scalar *mass* m_μ, can be located at the position $\mathbf{r}_\mu(t)$, and has the *velocity* $\mathbf{v}_\mu(t)$.

Integrated Quantum Hybrid Systems
Janik Wolters
Copyright © 2015 Pan Stanford Publishing Pte. Ltd.
ISBN 978-981-4463-82-9 (Hardcover), 978-981-4463-83-6 (eBook)
www.panstanford.com

The normal modes, also called *radiation* or simply *light*, are less intuitive objects, which are delocalized within the finite volume L^3 in which the experiment takes place. Each mode l can be characterized by a *wave vector* $\mathbf{k}_l$ and a *polarization* s_l and has the complex *amplitude* $\alpha_l(t)$. The wave vector is of the form

$$\mathbf{k} = \left(\frac{2\pi}{L}\right)^3 \mathbf{n}, \tag{2.1}$$

with $\mathbf{n}$ being a three-dimensional vector of integer numbers. For each $\mathbf{k}$ two possible polarizations $s = 1$, 2 exist, which are associated with two unit vectors $\epsilon_{1,\mathbf{k}}$ and $\epsilon_{2,\mathbf{k}}$, both being orthogonal to each other and to $\mathbf{k}$.

The normal modes can be related to the complex normal mode field $\mathbf{a}_l(\mathbf{r}, t)$. One particular choice of the normal mode field is

$$\mathbf{a}_l(\mathbf{r}, t) = \epsilon_l \alpha_l(t) e^{i\mathbf{k}\cdot\mathbf{r}}. \tag{2.2}$$

As investigated in Part III, other choices are also possible, and the only requirement on the normal mode fields is normalization and orthogonality in the sense of

$$\int_{L^3} d^3\mathbf{r}\, \mathbf{a}_l^*(\mathbf{r}, t) \mathbf{a}_m(\mathbf{r}, t) = \alpha_l^*(t)\alpha_m(t) L^3 \delta_{ml}. \tag{2.3}$$

In electrodynamics it is common to define two quantities being proportional to the real and imaginary part of the normal modes. These are the *vector potential* $\mathbf{A}(\mathbf{r}_\mu, t)$, defined by

$$\mathbf{A}(\mathbf{r}, t) = \sum_l \frac{\mathcal{E}_l}{\omega_l} \left[\mathbf{a}_l(\mathbf{r}, t) + \mathbf{a}_l^*(\mathbf{r}, t)\right]$$

$$= \sum_l \epsilon_l \frac{\mathcal{E}_l}{\omega_l} \left[\alpha_l(t) e^{i\mathbf{k}_l\cdot\mathbf{r}} + \alpha_l^*(t) e^{-i\mathbf{k}_l\cdot\mathbf{r}}\right], \tag{2.4}$$

and the *transverse electric field* $\mathbf{E}_T(\mathbf{r}_\mu, t)$, defined by

$$\mathbf{E}_T(\mathbf{r}, t) = \sum_l \mathcal{E}_l \left[i\mathbf{a}_l(\mathbf{r}, t) - i\mathbf{a}_l^*(\mathbf{r}, t)\right]$$

$$= \sum_l \epsilon_l \mathcal{E}_l \left[i\alpha_l(t) e^{i\mathbf{k}_l\cdot\mathbf{r}} - i\alpha_l^*(t) e^{-i\mathbf{k}_l\cdot\mathbf{r}}\right]. \tag{2.5}$$

In these definitions, the proportionality constant between normal mode amplitude $\mathbf{a}_m(\mathbf{r}, t)$ and the above-defined field, $\mathcal{E}_l$, is somehow arbitrary. Later on, to give the Hamiltonian and canonical variables a simpler form, the particular choice

$$\mathcal{E}_l = \sqrt{\frac{\hbar \omega_l}{2 \varepsilon_0 L^3}}, \tag{2.6}$$

with the *frequency* ω_l, test-wise defined by the *dispersion relation*

$$\omega_l = \frac{|\mathbf{k}_l|}{\sqrt{\mu_0 \varepsilon_0}} = c|\mathbf{k}_l|. \tag{2.7}$$

turns out to be convenient.

2.2 Classical Particle and Field Dynamics

The dynamics of the normal modes and particles are determined by the *Hamilton equations* and the *minimal coupling Hamiltonian*. From these, Maxwell's equations and the Newton–Lorentz equation can be derived.

2.2.1 *Canonical Variables*

Within the framework of the Hamilton formalism the particle μ and the modes are described by the joint canonical variables

$$\mathbf{q}_\mu(t) := (\mathbf{r}_\mu(t),\, Q_0(t),\, \ldots,\, Q_l(t),\, \ldots),$$
$$\mathbf{p}_\mu(t) := (\mathbf{p}_\mu(t),\, P_0(t),\, \ldots,\, P_l(t),\, \ldots). \tag{2.8}$$

The particle-related components of the joint canonical variables are given by the position $\mathbf{r}_\mu(t)$ and the *canonical momentum* $\mathbf{p}_\mu(t)$ defined by

$$\mathbf{p}_\mu = m_\mu \mathbf{v}_\mu + q_\mu \mathbf{A}(\mathbf{r}_\mu, t). \tag{2.9}$$

The field-related components Q_l and P_l are given by the real and imaginary components of the mode amplitudes:

$$Q_l = \sqrt{2\hbar}\, \mathrm{Re}(\alpha_l), \tag{2.10}$$
$$P_l = \sqrt{2\hbar}\, \mathrm{Im}(\alpha_l). \tag{2.11}$$

These canonical variables are up to a factor identical with the Fourier components of the already introduced vector potential $\tilde{\mathbf{A}}_l$

and the transverse electric field $\tilde{\mathbf{E}}_{T,l}$:

$$\tilde{\mathbf{A}}_l = \boldsymbol{\epsilon}_l \, Q_l \sqrt{\frac{1}{\omega_l \varepsilon_0 L^3}}, \tag{2.12}$$

$$\tilde{\mathbf{E}}_{T,l} = -\boldsymbol{\epsilon}_l \, P_l \sqrt{\frac{\omega_l}{\varepsilon_0 L^3}}. \tag{2.13}$$

2.2.2 Hamilton Equations

The dynamics of the i-th component of the joint canonical variables of particle μ and the electromagnetic field are given by the Hamilton equations:

$$\frac{d\mathsf{q}_{\mu,i}}{dt} = +\frac{\partial \mathcal{H}_\mu}{\partial \mathsf{p}_{\mu,i}}, \tag{2.14}$$

$$\frac{d\mathsf{p}_{\mu,i}}{dt} = -\frac{\partial \mathcal{H}_\mu}{\partial \mathsf{q}_{\mu,i}}. \tag{2.15}$$

Here, $\mathcal{H}_\mu$ is the *minimal coupling Hamiltonian* [11]

$$\mathcal{H}_\mu = \frac{1}{2m_\mu} \left[\mathbf{p}(t) - q\mathbf{A}(\mathbf{r}_\mu, t) \right]^2 + q\varphi(\mathbf{r}_\mu)$$
$$+ \sum_l \hbar\omega_l \left[Q_l^2(t) + P_l^2(t) \right], \tag{2.16}$$

where the *Coulomb potential* $\varphi(\mathbf{r}, t)$ is defined by

$$\varphi(\mathbf{r}, t) = \frac{1}{4\pi\varepsilon_0} \int d^3\mathbf{r}' \frac{\rho(\mathbf{r}')}{|\mathbf{r} - \mathbf{r}'|}. \tag{2.17}$$

This Coulomb potential gives the interaction energy of the charge μ with the *charge density*

$$\rho(\mathbf{r}, t) = \sum_{\nu \neq \mu} q_\nu \delta[\mathbf{r} - \mathbf{r}_\nu(t)] \tag{2.18}$$

generated by all other particles.

To get further insight into the behavior of the particles and fields, the Hamilton equations are explicitly evaluated and some remarks on the Coulomb potential follow.

2.2.3 *Coulomb Field*

The gradient of $\varphi(\mathbf{r}, t)$ is associated with the *Coulomb field*

$$\mathbf{E}_C(\mathbf{r}, t) = -\nabla\varphi(\mathbf{r}, t). \tag{2.19}$$

From Eqs. 2.17 and 2.19, one can deduce that the Coulomb field satisfies

$$\nabla \cdot \mathbf{E}_C(\mathbf{r}, t) = \frac{1}{\varepsilon_0}\rho(\mathbf{r}, t). \tag{2.20}$$

This can be Fourier-transformed to

$$i\mathbf{k_n}(t) \cdot \tilde{\mathbf{E}}_{C,\mathbf{n}}(t) = \frac{1}{\varepsilon_0}\tilde{\rho}_\mathbf{n}(t). \tag{2.21}$$

The Coulomb field is a purely *longitudinal* field, which is always parallel to the wave vector $\mathbf{k}$.

2.2.4 *Space-Related Variables*

The dynamics of the particle position variable $\mathbf{r}_\mu$ is determined by Eq. 2.14 in combination with the definition of the canonical momentum, Eq. 2.9:

$$\frac{d\mathbf{r}_\mu}{dt} = \mathbf{v}_\mu. \tag{2.22}$$

Hence, the previously introduced velocity is indeed the time derivative of the particle position, and thus the *electric current density* can be defined as

$$\mathbf{j}(\mathbf{r}, t) = \sum_\mu \mathbf{v}_\mu(t)q_\mu\delta[\mathbf{r} - \mathbf{r}_\mu(t)]. \tag{2.23}$$

Using Gauss's theorem, one can further identify

$$\nabla \cdot \mathbf{j}(\mathbf{r}, t) = \frac{d}{dt}\rho(\mathbf{r}, t). \tag{2.24}$$

This can be Fourier-transformed, and together with Eq. 2.21 it is found that the longitudinal currents $\tilde{\mathbf{j}}_{\parallel,l}(t)$ are responsible for changes of the Coulomb field:

$$\tilde{\mathbf{j}}_{\parallel,l}(t) = -\varepsilon_0\frac{d}{dt}\tilde{\mathbf{E}}_{C,l}(t). \tag{2.25}$$

The transverse components of the current $\tilde{\mathbf{j}}_{\perp,l}(t)$ which are orthogonal to the wave vector $\mathbf{k}$ are given by

$$\tilde{\mathbf{j}}_{\perp,l}(t) = \frac{1}{L^3} \sum_\mu q_\mu \mathbf{v}_\mu(t) \cdot \boldsymbol{\epsilon}_l \exp(-i\mathbf{k}_l \cdot \mathbf{r}_\mu). \tag{2.26}$$

Here, $\boldsymbol{\epsilon}_l$ and $\mathbf{k}_l$ are identical to those characterizing the normal mode l.

2.2.5 Radiation-Related Variables

To calculate the dynamics of the radiation-related variables $P_l(t)$ and $Q_l(t)$, the vector potential $\mathbf{A}(\mathbf{r}_\mu, t)$ in the Hamiltonian Eq. 2.16 must be expressed in terms of P_l and Q_l by using Eqs. 2.4, 2.10, and 2.11. The subsequent evaluation of Eqs. 2.14 and 2.15 yields the dynamics of the canonical variables, which are used to compute the dynamics of the normal modes, and the Fourier components of the vector potential as well as the transverse electric field, respectively. This results in

$$\frac{da_l}{dt} = -i\omega_l \alpha_l + \frac{i}{2\varepsilon_0 \mathcal{E}_l} \tilde{\mathbf{j}}_{\perp,l}(t), \tag{2.27}$$

$$\frac{d\tilde{\mathbf{A}}_l}{dt} = -\tilde{\mathbf{E}}_{T,l}(t), \tag{2.28}$$

$$\frac{d\tilde{\mathbf{E}}_{T,l}}{dt} = \omega_l^2 \tilde{\mathbf{A}}_l(t) - \frac{1}{\varepsilon_0} \tilde{\mathbf{j}}_{\perp,l}(t). \tag{2.29}$$

The normal mode amplitudes a_l behave like harmonic oscillators with frequency ω_l and are driven by the transverse currents.

2.2.6 Maxwell Equations

In the following, the dynamics of the so-called *electric field* defined by

$$\tilde{\mathbf{E}}_l(t) = \tilde{\mathbf{E}}_{T,l}(t) + \tilde{\mathbf{E}}_{C,l}(t), \tag{2.30}$$

and the *magnetic field* defined by

$$\tilde{\mathbf{B}}_l(t) = i\mathbf{k}_l \times \tilde{\mathbf{A}}_l(t). \tag{2.31}$$

are derived. First, the transversality of $\tilde{\mathbf{E}}_{T,l}(t)$ can be used to reformulate Eq. 2.21:

$$i\mathbf{k}_l(t) \cdot \tilde{\mathbf{E}}_l(t) = \frac{1}{\varepsilon_0} \tilde{\rho}_l(t). \tag{2.32}$$

Second, the magnetic field $\tilde{\mathbf{B}}_l(t)$ is purely transverse by definition and thus the relation

$$i\mathbf{k}_l \cdot \tilde{\mathbf{B}}_l(t) = 0 \tag{2.33}$$

holds. Furthermore, applying $i\mathbf{k}_l \times \ldots$ to both sides of Eq. 2.28 and using that $\mathbf{k}_l$ is parallel to the Coulomb field, results in

$$-i\mathbf{k}_l \times \tilde{\mathbf{E}}_l(t) = \frac{d}{dt} \left(i\mathbf{k}_l \times \tilde{\mathbf{A}}_l \right). \tag{2.34}$$

Similar, by adding Eqs. 2.25 and 2.29, using the transversality of $\tilde{\mathbf{A}}_l$ one finds

$$i\mathbf{k}_l \times \tilde{\mathbf{B}}_l(t) = \frac{1}{c^2}\frac{d}{dt}\tilde{\mathbf{E}}_l + \frac{1}{\varepsilon_0 c^2}\tilde{\mathbf{j}}_l(t). \tag{2.35}$$

Transforming from Fourier space to real space Eqs. 2.32, 2.33, 2.34, and 2.35 read

$$\nabla \times \mathbf{E}(\mathbf{r}, t) = \frac{1}{\varepsilon_0}\rho(\mathbf{r}, t), \tag{2.36}$$

$$\nabla \times \mathbf{B}(\mathbf{r}, t) = 0, \tag{2.37}$$

$$\nabla \times \mathbf{E}(\mathbf{r}, t) = -\dot{\mathbf{B}}(\mathbf{r}, t), \tag{2.38}$$

$$\nabla \times \mathbf{B}(\mathbf{r}, t) = \frac{1}{c^2}\dot{\mathbf{E}}(\mathbf{r}, t) + \frac{1}{\varepsilon_0 c^2}\mathbf{j}(\mathbf{r}, t). \tag{2.39}$$

This set of four partial differential equations is well known as *Maxwell equations*, which will be used in Part III to investigate the properties of optical microstructures.

2.2.7 Momentum-Related Variables

To obtain the particle dynamics Eq. 2.15 must be evaluated for the canonical momentum $\mathbf{p}_\mu(\mathbf{r}, t)$. For the x component of the momentum this results in

$$\frac{dp_{x,\mu}}{dt} = -q\,\mathbf{v}_\mu \cdot \frac{\partial \mathbf{A}}{\partial r_x} + q\frac{\partial \varphi}{\partial r_x}, \tag{2.40}$$

where the definition of the canonical momentum Eq. 2.9 is used. Independently, using Eq. 2.9 again, one finds

$$\frac{dp_{x,\mu}}{dt} = m\frac{d^2 r_{x,\mu}}{dt^2} + q\left(\nabla A_x \cdot \mathbf{v}_\mu + \frac{\partial \mathbf{A}}{\partial t}\right). \tag{2.41}$$

Combining Eqs. 2.40 and 2.41, the force acting on the charge μ can be found:

$$m\frac{d^2\mathbf{r}_\mu}{dt^2} = q\left(\nabla\varphi - \frac{\partial \mathbf{A}}{\partial t} + \mathbf{v}_\mu \times \nabla \times \mathbf{A}\right). \tag{2.42}$$

Using the definition of $\mathbf{B}$ (Eq. 2.31), the definition of the Coulomb field Eqs. 2.19, and 2.28, the well-known *Newton–Lorentz equation* is recovered:

$$m\frac{d^2\mathbf{r}_\mu}{dt^2} = q\left(\mathbf{E} + \mathbf{v}_\mu \times \mathbf{B}\right). \tag{2.43}$$

2.2.8 Dipole Approximation

In the following, the so-called *dipole approximation* is introduced. This widely used approximation allows to significantly simplify the Hamiltonian. Here, for simplicity, the index μ is suppressed, assuming that only the single particle μ with charge q is treated.

Gauge Invariance It is well known from classical electrodynamics [12] that the equations of motion for the fields $\mathbf{E}$ and $\mathbf{B}$ (Eqs. 2.36–2.39) and hence the particles (Eq. 2.43) are invariant under the gauge transformations

$$\mathbf{A}'(\mathbf{r}, t) = \mathbf{A} + \nabla \chi (\mathbf{r}, t), \tag{2.44}$$

$$\varphi'(\mathbf{r}, t) = \varphi(\mathbf{r}, t) - \frac{\partial}{\partial t} \chi (\mathbf{r}, t), \tag{2.45}$$

with an arbitrary gauge function $\chi (\mathbf{r}, t)$. For the discussion above, $\nabla \chi (\mathbf{r}, t) = 0$ was implicitly chosen. This *Coulomb gauge* makes the vector potential $\mathbf{A}(\mathbf{r}, t)$ purely transverse. Nevertheless, this is not always the best choice, as shown in the next paragraph.

Dipole Approximation In optics, the vector potential $\mathbf{A}(\mathbf{r}, t)$ varies on length scales of the *wavelength*

$$\lambda = \frac{2\pi}{|\mathbf{k}|} = \frac{2\pi c}{\omega} \approx 100 \text{ nm.} \tag{2.46}$$

In contrast, single quantum systems have the size of the *Bohr radius* $a_0 \sim 0.05$ nm in case of single atoms, or up to about ~ 20 nm for quantum dots. Hence in case of a single charge located in a quantum system at position $\mathbf{r}_0 = 0$, the vector potential can assumed to be constant in the vicinity of the particles position. Thus in the Hamiltonian Eq. 2.16 the vector potential $\mathbf{A}(\mathbf{r}, t)$ can be replaced by $\mathbf{A}(0, t)$.

In this approximation, also known as dipole approximation, the Hamiltonian Eq. 2.16 can be significantly simplified by using the *Göppert–Mayer gauge* [13]

$$\chi (\mathbf{r}, t) = -\mathbf{r} \cdot \mathbf{A}(0, t). \tag{2.47}$$

Using Eqs. 2.28, 2.44, and 2.45, one finds the new fields

$$\mathbf{A}'(0, t) = 0, \tag{2.48}$$

$$\varphi'(0, t) = \varphi(\mathbf{r}, t) - \mathbf{r} \cdot \mathbf{E}_T (0, t). \tag{2.49}$$

With this, the Hamiltonian 2.16 reads

$$\mathcal{H} = \frac{\mathbf{p}(t)^2}{2m} - q\mathbf{r} \cdot \mathbf{E}_T(0, t) + q\varphi(\mathbf{r})$$

$$+ \sum_l \hbar\omega_l \left[Q_l^2(t) + P_l^2(t) \right], \qquad (2.50)$$

The terms coupling the vector potential $\mathbf{A}$ and the momentum $\mathbf{p}$ vanish and a new term $-q\mathbf{r} \cdot \mathbf{E}_T$ appears. As $-q\mathbf{r}$ corresponds to the dipole moment of the particle, this coupling is called *dipole coupling*.

In this approximation the fields are assumed to be constant in the proximity of the considered particle, and hence higher multipole moments get neglected. In a more precise treatment using the *Power–Zienau–Wooley transformation*, this can be avoided [14].

2.3 The Quantized Hamiltonian

The whole plethora of classical electromagnetic phenomena can be derived from Maxwell's equations (Eqs. 2.36–2.39) and the Newton–Lorenz equation (Eq. 2.43). They represent a major part of the theoretical description of the physical world until the 20th century. Then E. Schrödinger [15] and W. Heisenberg [16] developed a new formalism, quantum mechanics. While this new theory precisely predicted the energy levels of atoms, the according transition rates by spontaneous emission remained unpredictable until P. Dirac applied the quantization formalism to the electromagnetic field as well [17].

As the normal modes α_l behave as uncoupled harmonic oscillators,[1] it is assumed that their quantization rules are according to the mechanical harmonic oscillator. Hence, the more or less axiomatic *canonical quantization* formalism can be followed: Hilbert space elements Ψ, which are called *wave function*, are introduced and the canonical variables $\mathbf{r}$, $\mathbf{p}$, Q_l, P_l and the Hamiltonian $\mathcal{H}$ are replaced by operators $\hat{\mathbf{r}}$, $\hat{\mathbf{p}}$, $\hat{Q}_l$, $\hat{P}_l$, and $\hat{\mathcal{H}}$ acting on Ψ. The dynamics of the wave function are determined by the Schrödinger equation

$$i\hbar \frac{\partial}{\partial t} |\Psi\rangle = \hat{\mathcal{H}} |\Psi\rangle, \qquad (2.51)$$

---

[1]Here, it is assumed that the coupling via the transverse current 2.27 is small compared with the frequency ω_l.

and the operators are postulated to obey the *canonical commutator relations*

$$[\hat{q}_{\mu,i}, \hat{p}_{\mu,j}] = i\hbar\delta_{i,j}, \tag{2.52}$$

$$[\hat{q}_{\mu,i}, \hat{q}_{\mu,i}] = [\hat{p}_{\mu,i}, \hat{p}_{\mu,i}] = 0. \tag{2.53}$$

With this the quantization is completed and all other operators can be constructed from the fundamental ones.

The first operators to construct are the operators $\hat{a}_l$ associated with the normal mode amplitude α_l. This is done by using the definitions of Q_l and P_l Eq. 2.11 and replacing the variables by the corresponding operators. Furthermore, using the fundamental commutators, it is easy to derive commutator relations for $\hat{a}_l$ and $\hat{a}_l^\dagger$:

$$[\hat{a}_l, \hat{a}_m^\dagger] = \delta_{l,m}. \tag{2.54}$$

The *Hamilton operator* corresponding to the Hamiltonian in dipole approximation Eq. 2.50 reads, when expressing $\hat{Q}_l$, $\hat{P}_l$ in terms of $\hat{a}_l$, $\hat{a}_l^\dagger$ and furthermore inserting the definition of the transvers electric field (Eq. 2.5),

$$\hat{\mathcal{H}} = \frac{1}{2m}\hat{\mathbf{p}}(t)^2 - iq\sum_l \hat{\mathbf{r}} \cdot \boldsymbol{\epsilon}_l \mathcal{E}_l \left[\hat{a}_l(t) - \hat{a}_l^\dagger(t)\right] + q\varphi(\mathbf{r})$$

$$+ \sum_l \hbar\omega_l \left(\hat{a}_l(t)\hat{a}_l^\dagger(t) + \frac{1}{2}\right). \tag{2.55}$$

Here again, for simplicity, index μ is suppressed and only the single particles μ with charge q are assumed, e.g., an electron is treated quantum mechanically. An instructive introduction to many-particle quantum mechanics can be found, e.g., in Refs. [14, 18].

The Hamiltonian Eq. 2.55 can be decomposed into the sum of *particle Hamiltonian, radiation Hamiltonian,* and *interaction Hamiltonian*

$$\hat{\mathcal{H}} = \hat{\mathcal{H}}_P + \hat{\mathcal{H}}_R + \hat{\mathcal{H}}_I, \tag{2.56}$$

where

$$\hat{\mathcal{H}}_P = \frac{\hat{\mathbf{p}}^2}{2m} + \varphi(\hat{\mathbf{r}}), \tag{2.57}$$

$$\hat{\mathcal{H}}_R = \sum_l \hbar\omega_l \left[\hat{a}_l^\dagger \hat{a}_l + \frac{1}{2}\right], \tag{2.58}$$

$$\hat{\mathcal{H}}_I = -i\sum_l \hat{\mathbf{D}} \cdot \boldsymbol{\epsilon}_l \mathcal{E}_l \left[\hat{a}_l(t) - \hat{a}_l^\dagger(t)\right], \tag{2.59}$$

and the definition of the *dipole operator* $\hat{\mathbf{D}} = q\hat{\mathbf{r}}$ is used in Eq. 2.59.

The first term, $\mathcal{H}_P$, depends only on the particles' position and momentum. This term is extensively treated in quantum mechanics textbooks and some special cases are examined in Part II. The second term, $\mathcal{H}_R$, is purely related to the radiation and formally equals a harmonic oscillator. For most problems these two terms dominate the dynamics, and the interaction Hamiltonian $\mathcal{H}_I$ can be treated as a perturbation on the eigenstates and eigenvalues of $\mathcal{H}_P$ and $\mathcal{H}_R$. Before treating this perturbation in section 4 in detail, some properties of the unperturbed radiation field are reviewed in Chapter 3.

Chapter 3

Properties of the Quantized Electromagnetic Field

In this chapter some properties of the unperturbed quantized electromagnetic field modes described by $\hat{\mathcal{H}}_R$ (Eq. 2.58) are investigated. As the operators for different modes l commute, and the Hamiltonian $\hat{\mathcal{H}}_R$ does not mix different modes, all modes can be treated as independent. Thus, for simplicity it is assumed that only a single mode l is occupied.

In the next section, some observables of the electromagnetic field modes, namely the electric field operator $\hat{\mathbf{E}}$ and the magnetic field operator $\hat{\mathbf{B}}$ are introduced. Later on, in Section 3.2 the eigenvectors of $\hat{\mathcal{H}}_R$ and the corresponding energy eigenvalues are treated, while the important class of coherent states is revised in Section 3.3. The last Section 3.4 of this chapter is dedicated to the density of photon states.

3.1 Field Observables

The classical Newton–Lorenz equation Eq. 2.43 gives a hint about how the electromagnetic field modes described by $\hat{\mathcal{H}}_R$ can

Integrated Quantum Hybrid Systems
Janik Wolters
Copyright © 2015 Pan Stanford Publishing Pte. Ltd.
ISBN 978-981-4463-82-9 (Hardcover), 978-981-4463-83-6 (eBook)
www.panstanford.com

be observed. This is via the force acting on charged particles. Nevertheless, in Eq. 2.43 the normal modes do not contribute directly, but in form of the transverse component of the electric field $\mathbf{E}_T$ and the magnetic field $\mathbf{B}$. To derive the operators $\hat{\mathbf{E}}$ and $\hat{\mathbf{B}}$, Eqs. 2.4, 2.11, 2.13, 2.31 and a Fourier transformation are used to find classical expressions for $\mathbf{E}_T$ and $\mathbf{B}$ depending on α and α^*.

Using the *correspondence principle*, the normal modes α, α^* are replaced by the operators a, $a^\dagger$, resulting in operators corresponding to the vector fields:

$$\hat{\mathbf{A}}(\mathbf{r}) = \sum_l \boldsymbol{\epsilon}_l \frac{\mathcal{E}_l}{\omega_l} e^{i\mathbf{k}_l \cdot \mathbf{r}} a_l + \text{h.c.}, \tag{3.1}$$

$$\hat{\mathbf{E}}(\mathbf{r}) = \sum_l i\boldsymbol{\epsilon}_l \mathcal{E}_l e^{i\mathbf{k}_l \cdot \mathbf{r}} \hat{a}_l + \text{h.c.} = \hat{\mathbf{E}}^{(+)}(\mathbf{r}) + \hat{\mathbf{E}}^{(-)}(\mathbf{r}), \tag{3.2}$$

$$\hat{\mathbf{B}}(\mathbf{r}) = \sum_l \frac{i\mathbf{k}_l \times \boldsymbol{\epsilon}_l}{\omega_l} \mathcal{E}_l e^{i\mathbf{k}_l \cdot \mathbf{r}} \hat{a}_l + \text{h.c.} \tag{3.3}$$

In Eq. 3.2, the operator $\hat{\mathbf{E}}(\mathbf{r})$ is decomposed into the non-Hermitian operators $\hat{\mathbf{E}}^{(+)}(\mathbf{r})$ and $\hat{\mathbf{E}}^{(-)}(\mathbf{r})$, which are conjugates of each other. Using this decomposition, one can further define the Hermitian *field intensity operator*

$$\hat{I} = \hat{\mathbf{E}}^{(+)}(\mathbf{r}) \cdot \hat{\mathbf{E}}^{(-)}(\mathbf{r}) = \sum_l \mathcal{E}_l^2 \hat{a}_l^\dagger \hat{a}_l. \tag{3.4}$$

With this, the mean intensity of a state $|\psi\rangle$ can be related to the radiation Hamiltonian $\hat{\mathcal{H}}_R$:

$$\langle \psi | \hat{I} | \psi \rangle = \sum_l \mathcal{E}_l^2 \langle \psi | \hat{a}_l^\dagger \hat{a}_l | \psi \rangle. \tag{3.5}$$

This is in agreement with the classical expectation that the intensity is proportional to the field energy.

After introducing these field observables, the next section is dedicated to the eigenvectors and eigenvalues of the Hamiltonian $\hat{\mathcal{H}}_R$.

3.2 Fock States

Within the canonical quantization the radiation Hamiltonian $\hat{\mathcal{H}}_R$ is constructed to equal the Hamiltonian of a set of independent *quantum harmonic oscillators.* Thus the eigenvalues and eigenstates $|n_l\rangle$ of $\hat{\mathcal{H}}_R$ can be directly adapted from quantum mechanics text books, e.g., Ref. [19]. Here these results are only briefly reviewed.

From the commutator relations Eq. 2.54 follows:

$$\hat{a}_l |n_l\rangle = \sqrt{n_l} |n_l - 1\rangle, \tag{3.6}$$

$$\hat{a}_l^\dagger |n_l\rangle = \sqrt{n_l + 1} |n_l + 1\rangle, \tag{3.7}$$

where n_l is the eigenvalue of the *number operator* [20, 21]

$$\hat{n}_l = \hat{a}_l^\dagger \hat{a}_l. \tag{3.8}$$

For these eigenvalues

$$\hat{n}_l |n_l\rangle = n_l |n_l\rangle, \quad n_l \in \mathbb{N}^+ \tag{3.9}$$

holds. The energy of the eigenstates of $\mathcal{H}_R$, also called *Fock states*, is

$$\langle n_l | \mathcal{H}_R | n_l \rangle = \hbar \omega_l \left(n_l + \frac{1}{2} \right). \tag{3.10}$$

In quantum electrodynamics the excitation quanta of the electromagnetic field are called photons, and hence n_l is identified as the photon number in the mode l and for the eigenstates of $\mathcal{H}_R$ the name *photon number states* is also justified. Applied to such an energy eigenstate, the operators $\hat{a}_l^\dagger$ and $\hat{a}_l$ increase and decrease the photon number by 1. Consequently, these operators are called *creation* and *annihilation operator*, respectively. When the annihilation operator is applied repetitively, one reaches the state $|0_l\rangle$ with the lowest eigenvalue $n_l = 0$, usually called *vacuum state*. Surprisingly, this state does not have zero energy, but from Eq. 3.11 one obtains

$$\langle 0 | \mathcal{H}_R | 0 \rangle = \sum_l \frac{1}{2} \hbar \omega_l. \tag{3.11}$$

For an infinite number of modes this expression diverges. Here, one major problem of quantum electrodynamics is found. It is commonly solved by energy renormalization, i.e., ignoring the vacuum state energy. Noticeably, from the vacuum state, all other states can be constructed by iterative application of the creation operator $\hat{a}_l^\dagger$:

$$|n_l\rangle = \frac{1}{\sqrt{n_l}} (\hat{a}_l^\dagger)^n |0\rangle. \tag{3.12}$$

The mean value of the electric field operator Eq. 3.2 associated with a Fock state is given by

$$\mathbf{E}_{n_l} = \langle n_l | \hat{E} | n_l \rangle = 0. \tag{3.13}$$

Surprisingly, the electric field vanishes. With the photon number operator, the intensity defined in Eq. 3.4 can be calculated. The intensity of a Fock state is given by

$$I_{n_l} = \langle n_l | \hat{I} | n_l \rangle = n_l \, \mathcal{E}_l^2. \tag{3.14}$$

This is nothing else but the photon number multiplied by the constant $\mathcal{E}_l^2$, which can be interpreted as electric field intensity per photon of mode l.

The Schrödinger equation Eq. 2.51 yields the temporal behavior of the state $|n_l(t)\rangle$:

$$|n_l(t)\rangle = e^{-i\omega_l(n_l+\frac{1}{2})t} |n_l(0)\rangle. \tag{3.15}$$

3.3 Coherent States

While the above section treated the eigenstates of the Hamiltonian of free radiation $\mathcal{H}_R$, this section treats the *quasi-classical states*.[1] These are constructed as eigenstates of the annihilation operator, and the associated electric field is a close approximation of the classical plane wave.

As a defining requirement it is postulated that

$$\hat{a}_l |\alpha_l\rangle = \alpha_l |\alpha_l\rangle \tag{3.16}$$

holds for the quasi-classical state $|\alpha_l\rangle$.

It is easy to prove [20], that this is fulfilled by a superposition of Fock states:

$$|\alpha_l\rangle = e^{-|\alpha_l|^2/2} \sum_{n=0}^{\infty} \frac{\alpha_l^n}{\sqrt{n!}} |n_l\rangle, \tag{3.17}$$

with arbitrary complex α_l. By projecting onto a specific number state $|n_l\rangle$ the probability of having a certain number of photons is calculated to follow the Poisson distribution:

$$P_{\alpha_l}(n_l) = | \langle \alpha_l | n_l \rangle |^2 = e^{-|\alpha_l|^2} \frac{|\alpha_l|^{2n}}{n!}. \tag{3.18}$$

---

[1]Sometimes this type of states are also referred to as *Glauber states*.

Thus, the photon number and energy in such a state are not fixed, as in the Fock states. When using the adjoint of Eq. 3.16 and the commutator relation Eq. 2.54, the expectation value of the photon number in $|\alpha_l\rangle$ is derived:

$$\langle \hat{n}_l \rangle = \langle \alpha_l | \hat{a}_l^\dagger \hat{a}_l | \alpha_l \rangle = |\alpha_l|^2. \tag{3.19}$$

Using this result and the definition of the intensity operator Eq. 3.4, this average intensity in the coherent field is given by

$$I_{\alpha_l} = \mathcal{E}_l^2 |\alpha_l|^2, \tag{3.20}$$

and equals the electric field intensity of the classical normal modes with amplitude α_l. As mentioned, the actually measured distribution of photon numbers fluctuates around the mean value. The distribution of these fluctuations has a variance of

$$\sigma_n = \langle \hat{n}_l^2 \rangle - \langle \hat{n}_l \rangle^2 = \left| \langle \alpha_l | \hat{a}_l^\dagger \hat{a}_l \hat{a}_l^\dagger \hat{a}_l | \alpha_l \rangle \right|^2 - |\alpha_l|^4 = |\alpha_l|^2. \tag{3.21}$$

Hence, the root-mean-square deviation $\sqrt{\sigma_n}$ of the average photon number is given by the square root of the photon number.

By using Eq. 3.15 the temporal evolution of the eigenvalue of the annihilation operator $\hat{a}_l$ is derived to be

$$\hat{a}_l \, |\alpha_l(t)\rangle = \alpha_l(t_0)e^{-i\omega_l t} \, |\alpha_l(t)\rangle. \tag{3.22}$$

During the time evolution the classical state remains a classical state, while the corresponding eigenvalues rotates with frequency ω_l in the complex plain. Consequently, the dynamics of the eigenvalues equals the dynamics of the classical normal modes in absence of external currents as described by Eq. 2.27. Furthermore, from this the dynamics of the expectation value for the electric field operator Eq. 3.2 are easily computed:

$$\langle \alpha_l(t)| \, \hat{\mathbf{E}}_T(\mathbf{r}) \, |\alpha_l(t)\rangle = i\epsilon_l \mathcal{E}_l e^{i\mathbf{k}_l \cdot \mathbf{r} - \omega_l t}\alpha_l(t_0) + \text{c.c.} \tag{3.23}$$

The electric field associated with the coherent state is a plane wave with wave vector $\mathbf{k}_l$ and frequency ω_l.

Such states are not just a theoretical concept, but for example generated by lasers and Hertzian dipoles. In Chapter 4 they will be used to describe several interesting cases of light–matter interaction.

3.4 Quasi Continuum and Density of States

When the quantization volume L is large, the spacing between the wave vectors $\mathbf{k}$ Eq. 2.1 will get very small. In this case, sums over l can be transformed into an integral:

$$\sum_l = \sum_{\mathbf{k},s} \to \frac{V}{(2\pi)^3} \sum_s \int d^3\mathbf{k}. \tag{3.24}$$

Transforming this into spherical coordinates by making use of the dispersion relation Eq. 2.7 and the photon energy $E = \hbar\omega$ this gets

$$\sum_l \to \frac{V}{(2\pi\hbar c)^3} \sum_s \int \int \int d\varphi \, d\theta \, dE \, E^2 \sin(\theta), \tag{3.25}$$

where the *density of states* at a certain energy E, with the k vector pointing into a certain direction given by θ and φ can be identified as

$$\rho(\varphi, \theta, E) = \frac{V}{(2\pi\hbar c)^3} E^2 \sin(\theta). \tag{3.26}$$

As the photon energy was used to derive this expression, the density of states can be regarded as a purely quantum mechanical expression. In the next chapter, this expression will be used to calculate the total spontaneous emission rate of single emitters from Fermi's golden rule.

Chapter 4

Light–Matter Interaction

The Hamiltonians of the particle $\mathcal{H}_P$ and radiation $\mathcal{H}_R$ commute and thus have a common set of eigenvectors. These are given as the tensor product of the electronic state $|a\rangle$ and the eigenstates $|n_l\rangle$ of the electromagnetic modes:

$$|\phi\rangle = |a\rangle \otimes |n_1\rangle \otimes \cdots \otimes |n_l\rangle \otimes \ldots = |a, n_1, \ldots, n_l \ldots\rangle. \quad (4.1)$$

In this section on light–matter interaction the Hamiltonian $\mathcal{H}_I$ is treated as a perturbation on these eigenstates $|\phi\rangle$. Under the influence of this perturbation the states $|\phi\rangle$ are no longer eigenstates and thus get explicitly time dependent. This results in a plethora of phenomena, like absorption, spontaneous emission, Rabi oscillations, and many more.

4.1 Second-Order Perturbation Theory

Here, the interaction of an two-level electronic system described by the Hamiltonian $\mathcal{H}_E$ with eigenstates $|i\rangle, |f\rangle$ and a photon Fock state $|n\rangle$ in the single mode l is considered. In second-order perturbation theory *Fermi's golden rule* gives the transition rate from initial state $|1\rangle = |i, n_i\rangle$ to final state $|2\rangle = |f, n_f\rangle$, when the energies E of both

Integrated Quantum Hybrid Systems
Janik Wolters
Copyright © 2015 Pan Stanford Publishing Pte. Ltd.
ISBN 978-981-4463-82-9 (Hardcover), 978-981-4463-83-6 (eBook)
www.panstanford.com

states are equal:

$$\Gamma_{1\to2} = \frac{2\pi}{\hbar}\rho(E)|\langle 2|\mathcal{H}_I|1\rangle|^2, \tag{4.2}$$

with the density of final states $\rho(E)$. Using the definition of $\mathcal{H}_I$ Eqs. 2.59, 3.6, and 3.7 this can be directly evaluated:

$$\Gamma_{1\to2} = \frac{2\pi\,\mathcal{E}_l^2}{\hbar}\rho(E)\left|\langle f|\hat{\mathbf{d}}\cdot\boldsymbol{\epsilon}_l|i\rangle\right|^2 \left|\langle n_f|\hat{a}_l|n_i\rangle\right|^2 \tag{4.3}$$

$$+ \frac{2\pi\,\mathcal{E}_l^2}{\hbar}\rho(E)\left|\langle f|\hat{\mathbf{d}}\cdot\boldsymbol{\epsilon}_l|i\rangle\right|^2 \left|\langle n_f|\hat{a}_l^\dagger|n_i\rangle\right|^2.$$

Since Fock states with different photon numbers are orthogonal, the first part contributes only when the final photon state $|n_f\rangle$ contains one photon less than the initial state $|n_i\rangle$. This term corresponds to the *absorption* of one photon. In contrast, the second term corresponds to the *emission* of a single photon into mode l.

4.1.1 *Absorbtion*

In the case of photon absorption, the total photon energy $\langle n|\mathcal{H}_L|n\rangle$ in the final state is $\hbar\omega_l$ less than in the initial state. To fulfill the energy resonance condition of Fermi's golden rule, this must be compensated by the particle system. Hence, for the particle energies $E_i = \langle i|\mathcal{H}_P|i\rangle$ it is required that

$$E_2 - E_1 = \hbar\omega_l. \tag{4.4}$$

The energy of the final particle state must be $\hbar\omega_l$ higher than the one of its initial state. Under this absorption resonance condition the rate Eq. 4.3 simplifies, when using the intensity of the light field I_{n_l} defined in Eq. 3.14, to

$$\Gamma_{1\to2} = \frac{2\pi}{\hbar}I_{n_l}\rho(E)\left|\langle f|\hat{\mathbf{D}}\cdot\boldsymbol{\epsilon}_l|i\rangle\right|^2 = \gamma_{exc}. \tag{4.5}$$

As expected from semiclassical theory, the transition rate from particle state $|i\rangle$ to $|f\rangle$ is proportional to the intensity of the perturbing electromagnetic field and the dipole matrix element.

4.1.2 *Emission*

Emission of single photons corresponds to the second term in Eq. 4.3. Here the resonance condition gives

$$E_2 - E_1 = -\hbar\omega_l. \tag{4.6}$$

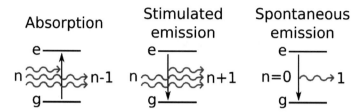

Figure 4.1 Schematic illustration of the absorption and emission processes. Absorption: Under the influence of the n photon field the two-level system is transferred from the ground state g to the excited state e, while one photon is absorbed. Stimulated emission: Under the influence of the n photon field the two-level system is de-excited into the ground state g under emission of one photon. Spontaneous emission: Under the influence of the vacuum field the two-level system is de-excited into the ground state g under emission of one photon.

For photon emission, the energy of the initial particle state must exceed that of its final state by the photon energy $\hbar\omega_l$. Analogous to the previous case of absorption, the rate Eq. 4.3 simplifies to

$$
\begin{aligned}
\Gamma_{1 \to 2} &= \frac{2\pi}{\hbar} I_{n_l} \rho(E) \left| \langle f | \hat{\mathbf{D}} \cdot \boldsymbol{\epsilon}_l | i \rangle \right|^2 \\
&+ \frac{2\pi}{\hbar} \mathcal{E}_l^2 \rho(E) \left| \langle f | \hat{\mathbf{D}} \cdot \boldsymbol{\epsilon}_l | i \rangle \right|^2 = \gamma_{ind} + \Gamma_{spon}.
\end{aligned}
\tag{4.7}
$$

The first term γ_{ind} is identical to the absorption rate Eq. 4.5. It is interpreted as *stimulated emission*. The photon field with intensity I_{n_l} induces oscillations of the particles dipole and thereby stimulates the de-excitation. This result can also be reproduced by semiclassical theories, in which the electromagnetic field is treated as a classical quantity. In contrast, the second term Γ_{spon} being proportional to $\mathcal{E}_l^2$ is purely a consequence of the commutator relations Eqs. 2.52 and 2.53, i.e., the quantization of the electromagnetic field. This term is called *spontaneous emission* and allows transitions even when the photon mode is initially in the vacuum state.

As here $\mathcal{E}_l^2$ takes the role of the intensity in the stimulated process, it is interpreted as the *vacuum field intensity*.

4.1.3 *Photon Detection and Statistics*

The above-described photon absorption mechanism is particularly useful to detect light. Indeed, apart from a few experiments on non-demolition measurements [22, 23], all photon detectors are based on absorption. In a detector single photons are absorbed and thereby change the state of a single particle, usually an electron [24]. Now, it is hardly possible to detect a single elementary charge and hence the signal has to be amplified prior to the final registration by classical sensing devices like ammeters. For example, in an avalanche photodiode (APD), an electron is brought from the valence to the conduction band. There, an externally applied voltage generates a strong field that accelerates the electron and thereby triggers a macroscopic electron avalanche which can be measured as a classical macroscopic current.

Single-Photon Detection Probability While the absorption of a present photon happens with almost unity probability within a short time interval Δt, the avalanche is triggered only with probability η. Thus, when the initial photon state is $|\phi\rangle = \sum_n c_n |n\rangle$, the number of expected photon detection events within the interval Δt can be obtained by summing up Eq. 4.3 over all possible final photon states and integrating the rate over Δt:

$$N_{det} = \eta \sum_{n'} \left| \langle n'|\hat{a}|\phi\rangle \right|^2 = \eta \langle \phi|\hat{a}^\dagger \hat{a}|\phi\rangle = \eta \langle n\rangle. \qquad (4.8)$$

As one might expect, the number of detected photons is proportional to the average number of photons times the quantum efficiency η.

Two-Photon Detection Probability After the detection event, an observer knows for sure that at least one photon was present and the new wave function contains one photon less. Thus, the new (unnormalized) wave function can be written as

$$|\tilde{\phi}'\rangle = \sum_n c_n \sqrt{n} |n - 1\rangle, \qquad (4.9)$$

where the factor of $\sqrt{n}$ accounts for the increased detection probability for higher photon number states. This new wave function can also be constructed by letting the annihilation operator

$\hat{a}(t)$ act at time t and hence the normalized new wave function can be written as

$$|\phi'\rangle = \frac{\hat{a}(t)\,|\phi\rangle}{\sqrt{\langle\phi|\hat{a}^\dagger(t)\hat{a}(t)|\phi\rangle}}. \tag{4.10}$$

Inserting this into Eq. 4.8, the probability to detect a second photon at time $t + \tau$ conditioned on the detection of a photon at time t is found to be

$$N(t + \tau|t) = \frac{\eta}{\langle n(t)\rangle}\,\langle\phi'|\hat{a}^\dagger(t+\tau)\hat{a}(t+\tau)|\phi'\rangle. \tag{4.11}$$

By multiplying this with the probability of the photon detection at time t, the joint two-photon detection probability, i.e., the unnormalized second-order correlation function $G^{(2)}(t, t + \tau)$ is found:

$$G^{(2)}(t, t + \tau) = \eta^2\,\langle\phi|\hat{a}^\dagger(t)a^\dagger(t+\tau)a(t+\tau)\hat{a}(t)|\phi\rangle. \tag{4.12}$$

When this is normalized by the average intensities N_{det} the normalized second-order correlation function [25] is obtained:

$$g^{(2)}(t, t + \tau) = \frac{\langle\phi|\hat{a}^\dagger(t)a^\dagger(t+\tau)a(t+\tau)\hat{a}(t)|\phi\rangle}{\langle\phi|\hat{a}^\dagger(t)\hat{a}(t)|\phi\rangle\,\langle\phi|\hat{a}^\dagger(t+\tau)\hat{a}(t+\tau)|\phi\rangle}. \tag{4.13}$$

In the case of coherent states, which are eigenstates of the annihilation operator, one can simply calculate

$$g^{(2)}_{coherent}(t, t + \tau) = 1. \tag{4.14}$$

In contrast, in the case of Fock states the value of the $g^{(2)}$ function is given by

$$g^{(2)}_n(t, t + \tau) = \frac{n - 1}{n}. \tag{4.15}$$

This function has a value of zero when the initial state contains fewer than two photons. Remarkably for Fock states with large photon numbers, i.e., in the limit of Eq. 4.15 for $n \to \infty$, one obtains $g^{(2)}(t, t + \tau) = 1$, as for coherent states.

In the interesting cases where the number of photons changes with time, for example when an emitter is present, the calculation of the $g^{(2)}$ function requires some more work, as shown later on.

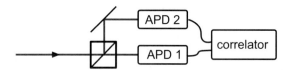

Figure 4.2 Schematic illustration of a Hanbury Brown and Twiss setup. The incoming photon state is split on a 50/50 beam splitter into two modes, each detected by a detector, e.g., an APD. A correlation electronic detects when both detectors click simultaneously, indicating a two-photon event. This allows to measure the autocorrelation function $g^{(2)}(t, t + \tau)$ even if the detectors have a long dead-time.

Hanbury Brown and Twiss Setup All detectors have a more or less short *dead-time* after the detection of a photon. For standard APDs, this time, in which no photons can be detected, is on the order of 50 ns. Furthermore, as photon number resolving detectors are not yet commercially available, it is rarely possible to directly measure two-photon events, i.e., the $g^{(2)}$ function, with one single detector [26].

To overcome this problem, usually a *Hanbury Brown and Twiss setup* (HBT) [27] consisting of two detectors and a 50/50 beam splitter is used. At the beam splitter the photon wave function is split up, giving equal detection probabilities at each detector. When only one photon is present, only one detector can click. In contrast when several photons are present, it becomes possible that both detectors click simultaneously. Using fast correlation electronics such events can be registered and compared to the single photon detection rate to obtain the autocorrelation function $g^{(2)}(t, t + \tau)$.

4.1.4 *Excitation of Two-Level Systems*

With the absorption and emission mechanism from the previous section, it is possible to describe a *two-level system* under the influence of an external photon field and the vacuum fluctuations. For experiments, the case of a coherent external single-mode photon field is particularly important as these can be generated by lasers. In the following paragraph, first the classical case is treated. There, spontaneous emission is neglected, while an analysis of the more interesting case with spontaneous emission follows.

Two-Level System in a Strong Laser Field   When only a single mode l which is dominated by a strong laser is considered, the photon field remains roughly unaffected by the two-level system and thereby can be eliminated from the dynamics. Furthermore, the density of states in the transition rates can be replaced by a Dirac delta. This case is illustrated in Fig. 4.3(a). The two-level system ground state occupation $\langle \hat{\rho}_g \rangle = \langle \Psi | g \rangle \langle g | \Psi \rangle$ and the excited state occupation $\langle \hat{\rho}_e \rangle = \langle \Psi | e \rangle \langle e | \Psi \rangle$ are completely described by the rate equations

$$\frac{d}{dt} \langle \hat{\rho}_e \rangle = \gamma \left(\langle \hat{\rho}_g \rangle - \langle \hat{\rho}_e \rangle \right), \tag{4.16}$$

$$\frac{d}{dt} \langle \hat{\rho}_g \rangle = \gamma \left(\langle \hat{\rho}_e \rangle - \langle \hat{\rho}_g \rangle \right). \tag{4.17}$$

Herein it is considered that according to Eqs. 4.5 and 4.7, the excitation rate γ_{exc} and the stimulated emission rates γ_{stim} are equal and thus one can define $\gamma = \gamma_{exc} = \gamma_{stim}$. The steady state of these dynamics is $\langle \hat{\rho}_g \rangle = \langle \hat{\rho}_e \rangle$, when absorption and spontaneous emission exactly compensate. Thus, in second-order perturbation theory, a two-level system can never be driven completely into the excited state.

Phonon Side Bands   In solid quantum systems, usually so-called *phonon side bands* (PSBs) exist. These provide levels $|e'\rangle$ above the excited state $|e\rangle$, from where nonradiative relaxation γ_{non} into the level $|e\rangle$ occurs. If this nonradiative relaxation is much faster than the stimulated emission process, which is usually the case for phonon processes occurring on ps timescales, the latter can be neglected. Thus, via this auxiliary level, stimulated emission can be prevented

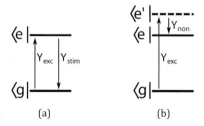

(a)   (b)

Figure 4.3   The two-level system without spontaneous emission. (a) The pure two-level system. (b) The two-level system with phonon sublevel $|e'\rangle$ and nonradiative relaxation rate γ_{non}.

and the excited state can be completely occupied with the rate

$$\gamma_{exc'} = \left(\frac{1}{\gamma_{exc}} + \frac{1}{\gamma_{non}} \right)^{-1}. \qquad (4.18)$$

In the considered limit of $\gamma_{non} \gg \gamma_{exc}$ this new rate $\gamma_{exc'}$ equals the original excitation rate γ_{exc}.

Spontaneous Emission   Another consequence of nonradiative decay is dissipation of energy into the phonon bath. Hence the excitation frequency can be above the frequency of the radiative decay. In frequency space the emitted photons can be easily separated from the strong excitation laser in mode l by using adequate filters. In this case, the spontaneously emitted photons in mode m might be detected by sensitive detectors, and it is worth having a closer look at this photon mode m.

It is assumed that at time t_0 the two-level system is in the ground state $|g\rangle$ and the photon mode m is in the vacuum state $|n_m = 0\rangle$. From this state $|g, 0\rangle$ the system is driven with the excitation rate γ_{exc} Eq. 4.5 to the state $|e, 0\rangle$. This state itself couples with the spontaneous emission rate γ_{spon} to $|g, 1\rangle$ (Eq. 4.7), from where the system is reexcited to $|e, 1\rangle$. From there spontaneous emission will take place again, at this time into $|g, 2\rangle$, and so on. The occupation of the mode m is steadily growing, as illustrated in Fig. 4.4. In this case an infinite number of states is necessary to describe the experiment.

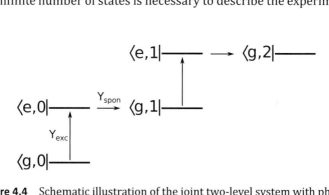

Figure 4.4   Schematic illustration of the joint two-level system with photon mode m, being initially empty. From state $|g, 0\rangle$ the system gets excited to the state $|e, 0\rangle$ and decays to $|g, 1\rangle$, and so on. If undamped, the photon number will grow steadily and an infinite number of states is needed to describe the system's dynamics.

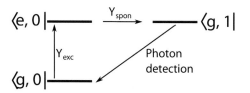

Figure 4.5 Absorption and emission process with photon damping. When assuming that the photons are immediately absorbed, only the states are necessary for a complete description of the dynamics.

Nevertheless, in most realistic experiments, the photons of mode m will be more or less immediately absorbed by the boundaries or detectors. Hence the photon number is always close to zero. This closes the system, as illustrated in Fig. 4.5, and only a finite set of differential equations is necessary to describe the situation:

$$\frac{d}{dt}\langle \hat{\rho}_{e,0} \rangle = \gamma_{exc}\langle \hat{\rho}_{g,0} \rangle - \gamma_{spon}\langle \hat{\rho}_{e,0} \rangle), \tag{4.19}$$

$$\frac{d}{dt}\langle \hat{\rho}_{g,0} \rangle = -\gamma_{exc}\langle \hat{\rho}_{g,0} \rangle + \gamma_{spon}\langle \hat{\rho}_{e,0} \rangle), \tag{4.20}$$

$$\langle \hat{\rho}_{g,1} \rangle = -\gamma_{spon}\langle \hat{\rho}_{e,0} \rangle \, dt. \tag{4.21}$$

Assuming that photons are detected with efficiency η, one obtains an average photon count rate proportional to the occupation of the upper level:

$$\langle N_{det} \rangle = \eta \, \langle \hat{\rho}_e \rangle \, \gamma_{spon}. \tag{4.22}$$

4.1.5 *Total Spontaneous Emission Rate*

Apparently the assumption of only one single mode is not justified for spontaneous emission, which might occur on any mode with the frequency $\omega = \omega_m$. Hence, with the density of states given by Eq. 3.26, the total spontaneous emission rate is given by integrating Γ_{spon} over all possible emission angles:

$$\begin{aligned}
\gamma_{spon} &= \frac{V E^2 \mathcal{E}^2}{4\pi^2 c^3 \hbar^4}\left|\langle f|\hat{\mathbf{D}}|i\rangle\right|^2 \iint d\varphi \, d\theta \, \sin^3(\theta) \\
&= \frac{\omega^3}{3\pi \varepsilon_0 \hbar c^3}\left|\langle f|\hat{\mathbf{D}}|i\rangle\right|^2.
\end{aligned} \tag{4.23}$$

Here, without loss of generality, one polarization ϵ_1 is chosen to be orthogonal to $\hat{\mathbf{D}}$, while the other polarization ϵ_2 encloses the angle θ, i.e., $\hat{\mathbf{D}} \cdot \epsilon_1 = \hat{D} \sin \theta$.

4.1.6 Steady State Solution of the Two-Level System

In the steady state, spontaneous emission exactly compensates excitation and the excited state occupation is given by

$$\langle \hat{\rho}_{ee} \rangle = \frac{\gamma_{exc}}{\gamma_{exc} + \gamma_{spon}}. \tag{4.24}$$

In this case Eq. 4.22 gets

$$\langle N_{det} \rangle = \eta \frac{\gamma_{exc} \gamma_{spon}}{\gamma_{exc} + \gamma_{spon}}. \tag{4.25}$$

Expressing γ_{exc} as a linear function of the intensity of the excitation light $I = \langle \hat{I} \rangle$ this can be rewritten as

$$N_{det}(I) = \frac{N_{inf} I}{I_{sat} + I}, \tag{4.26}$$

with the count rate at infinite excitation intensity N_{inf} and the saturation intensity I_{sat}, at which $N_{det} = N_{inf}/2$.

Remarkably, only in the low excitation limit the count rate depends linearly on the excitation power (cf. Fig. 4.6). Such behavior is not expected for classical emitters and a clear signature of individual quantum systems. Furthermore, Eq. 4.26 gives a limit for the maximal obtainable count rate, and sensitive detectors might be needed to observe the emission of individual quantum systems.

Saturation with Background Even worse, solid state emitters are usually embedded in some matrix which can be seen as an ensemble of many weakly excited quantum emitters. These generate a photon background depending linearly on the excitation power I. Furthermore, all detectors have a certain dark count rate N_{dark}. Thus Eq. 4.26 must be modified to

$$N_{det}(I) = \frac{N_{inf} I}{I_{sat} + I} + \eta a_{bg} I + N_{dark}. \tag{4.27}$$

The background and dark counts limit the usable excitation power to observe the individual two-level system, as illustrated in Fig. 4.6(b), or in worst case may even completely cover the two-level system emission.

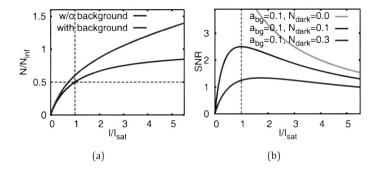

(a)   (b)

Figure 4.6 (a) The saturation behavior of two-level system emission. Fluorescence intensity of a pure two-level system and a two-level system with background of strength $a_{bg} = 0.1 \cdot N_{inf}$. Signal-to-noise ratio (SNR) for different dark count rates at constant background of strength $a_{bg} = 0.1 \cdot N_{inf}$.

4.1.7 Dynamic Behavior of Two-Level Systems

In this section, the dynamic behavior of the two-level system is investigated. At first, the case when the external laser field is suddenly turned off is investigated in the next paragraph, while the switch-on case is considered later on.

Fluorescence Lifetime After illumination for some time the two-level system can be considered to be in the steady state, with an excited state occupation given by Eq. 4.24. When now the excitation laser is suddenly switched off at time t_0, the excitation rate vanishes and Eq. 4.19 describes an exponential decay of the excited state into the ground state:

$$\rho_e(t) = \rho_e(t_0) \exp\left[-\gamma_{spon}(t - t_0)\right]. \tag{4.28}$$

According to Eq. 4.22 the probability of detecting a photon follows this process (Fig. 4.7). Hence, the spontaneous emission rate $\gamma_{spon} = 1/\tau_{soon}$ can be obtained from measurements of the time-resolved photon arrival probability after switch-off of the laser.

It is not necessary that the two-level system has reached the steady state when switching off the laser. The above holds also when excitation is only for a very short time. Thus pulsed lasers which provide a sufficiently fast slope are commonly used for such *fluorescence lifetime* measurements. When the fluorescence lifetime τ_{spon} of the studied two-level system is longer than the one of the

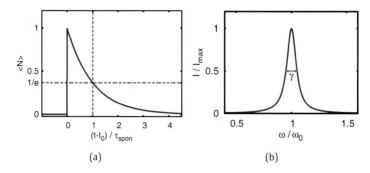

Figure 4.7 (a) The time-dependent photon detection probability of a two-level system with fluorescence lifetime τ_{spon} when excited with a short laser pulse at $t = t_0$. (b) The resulting spectrum of the emitted photon wave packets.

background, such time-resolved measurements are also useful to distinguish between background emission and the emission of the two-level system.

Time–Energy Uncertainity Obviously, when exciting the two-level system with a short laser pulse, the amplitude of the emitted photon is not constant in time, but the shape of its wave packet follows the exponential decay shown in Fig. 4.7. Thus, the emitted photon cannot be in an eigenstate of the Hamiltonian $\mathcal{H}_R$, which has a time evolution given by Eq. 3.15. However, the desired single photon wave packet $|\psi\rangle$ can be constructed as a superposition of one photon being in different modes [11]:

$$|\psi(t)\rangle = \sum_l c_l \exp(-i\omega_l t)\, |0, \ldots, n_l = 1, 0, \ldots\rangle, \quad (4.29)$$

with coefficients following a Lorentz distribution:

$$c_l = \frac{K}{\omega_l - \omega_0 + i\gamma_{spon}/2}, \quad (4.30)$$

with normalization constant $K = \sqrt{\gamma_{spon}c/L}$ and ω_0 corresponding to the resonance frequency according to Eq. 4.6. The absolute square of the coefficients c_l gives the intensity spectrum, corresponding to a Lorentzian of width γ_{spon}, centered at ω_0 (Fig. 4.7(b)):

$$|c_l|^2 = \frac{K^2}{(\omega_l - \omega_0)^2 + (\gamma_{spon}/2)^2}. \quad (4.31)$$

This behavior is a property of all energy levels which have a finite lifetime. The relation between spectral width $\Delta\omega$ and lifetime τ

$$\Delta\omega = 1/\tau \qquad (4.32)$$

can be multiplied by $\hbar$ and when interpreting τ as the uncertainty of the photon arrival time $\Delta t = \tau/2$ this gives the *time–energy uncertainty relation*

$$\Delta E \, \Delta t = \hbar/2. \qquad (4.33)$$

Although this is formally equal to the *Heisenberg uncertainty relation*, it is a pure consequence of the Fourier decomposition Eq. 4.29 of the exponential decay and not due to the commutator relation between noncommuting variables.

Switch-on   When assuming the two-level system initially in the state $|g, 0\rangle$ and suddenly switching the laser on, the system approximates the equilibrium with a time constant given by $\gamma_{exc} + \gamma_{spon}$. Exactly, by solving Eq. 4.21 one finds

$$\rho_e = \rho_{ss} \left(1 - e^{-(\gamma_{exc} + \gamma_{spon})t} \right), \qquad (4.34)$$

with the steady state occupation ρ_{ss} given by Eq. 4.24. This switch-on behavior becomes important when the statistics of the emitted photons are discussed in the next section.

4.1.8 *Photon Statistics of the Emission of Two-Level Systems*

In this section, the time-dependent second-order autocorrelation function Eq. 4.13 of the photon emitted by a single two-level system is calculated.

As discussed above, in the steady state the two-level system with one photon mode can be described by the wave function

$$|\psi\rangle = c_0 \, |g, 0\rangle + c_1 \, |e, 0\rangle + c_2 \, |g, 1\rangle. \qquad (4.35)$$

A photon detection event at time t_1 can be described by the annihilation operator. Thus, after detection, the wave function reduces to

$$\hat{a}(t_1) \, |\psi\rangle = c_2 \, |g, 0\rangle. \qquad (4.36)$$

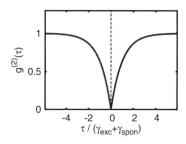

Figure 4.8 The $g^{(2)}$ function of a two-level system according to Eq. 4.38. At zero time delay τ the value is zero and follows the switch-on dynamics of the of the upper level Eq. 4.34.

This state evolves as described in the paragraph above and one finds for the detection of a second photon at time $t + \tau$

$$\hat{a}(t_1 + \tau)\hat{a}(t_1)\,|\psi\rangle = N_{det}\eta\rho_{ss}\left(1 - e^{-(\gamma_{exc}+\gamma_{spon})\tau}\right). \quad (4.37)$$

This can be normalized by the average photon count rate to get the normalized autocorrelation function $g^{(2)}(t, t + \tau)$ Eq. 4.13:

$$g^{(2)}(t, t + \tau) = \left(1 - e^{-(\gamma_{exc}+\gamma_{spon})\tau}\right), \quad (4.38)$$

illustrated in Fig. 4.8. For $\tau = 0$ the value of the autocorrelation function $g^{(2)}(t, t + \tau)$ is zero in accordance with the assumption that a single-photon Fock state is generated. For long time delay τ the $g^{(2)}$ function reaches 1, indicating that photons are uncorrelated.

4.1.9 Three-Level Systems

The three-level system is another simple model which is realized in many quantum systems like molecules or defects in diamond. As depicted in Fig. 4.9, such a system consists of a ground and an excited state (g and e) and an additional shelving state denoted by s. Usually, the latter one can be reached via a nonradiative transition with rate γ_{non} from the excited state and decays with the deshelving rate γ_{desh} into the ground state. Thus, the level dynamics are described by

$$\frac{d}{dt}\rho_{gg} = -\gamma_{exc}\rho_{gg} + \gamma_{spon}\rho_{ee} + \gamma_{desh}\rho_{ss}, \quad (4.39)$$

$$\frac{d}{dt}\rho_{ee} = -\left(\gamma_{spon} + \gamma_{non}\right)\rho_{ee} + \gamma_{exc}\rho_{gg}, \quad (4.40)$$

$$\frac{d}{dt}\rho_{ss} = -\gamma_{desh}\rho_{ss} + \gamma_{non}\rho_{ee}. \quad (4.41)$$

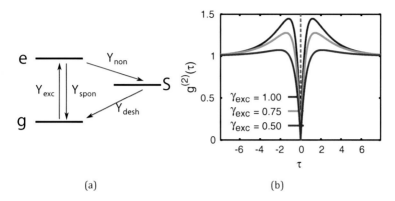

(a)   (b)

Figure 4.9 (a) Level structure and transition rates of the three-level system. (b) The autocorrelation function $g^{(2)(t,t+\tau)}$ of a three-level system according to Eq. 4.44 for $\gamma_{non} = 0.7$, $\gamma_{desh} = 0.2$ and $\gamma_{spon} = 1$ for different excitation rates γ_{exc}. While there is strong anti-bunching for $\tau = 0$ the $g^{(2)}$ function exhibits bunching for medium times, indicating an increased photon detection probability. The strength of this bunching depends on the excitation rate.

An analytical solution of this set of differential equations is given in Ref. [28] and discussed in the following.

The existence of the deshelving state has several consequences for the photon emission via spontaneous emission. First, the excited state lifetime is reduced to

$$\tau_e = \frac{1}{\gamma_{spon} + \gamma_{non}}. \tag{4.42}$$

Second, the spontaneous emission photon yield is reduced by the ratio between spontaneous emission rate γ_{soon} and nonradiative rate γ_{non}. In the worst case the nonradiative process may completely dominate the dynamics, making the system unobservable. For the maximal number of detected photons, one finds

$$N_{inf} = \frac{\gamma_{desh} + 1/\tau_e}{2\gamma_{exc} + \gamma_{non}}. \tag{4.43}$$

Nevertheless, the saturation behavior equals the two-level system and can be described by Eq. 4.27. When the deshelving rate is comparably low, this can furthermore reduce the photon emission rate of the system, as the shelving state is optically inactive. This has also important consequences on the switch-on dynamics and thus the second-order autocorrelation of the emitted light. The latter one is given by

$$g^{(2)}(t_0, t_0 + \tau) = 1 - (K + 1)e^{k_+\tau} + Ke^{k_-\tau}, \qquad (4.44)$$

with the coefficients $k_\pm = -\frac{1}{2}P \pm \sqrt{\frac{1}{4}P^2 - Q}$, where $P = \gamma_{exc} + \gamma_{spon} + \gamma_{non} + \gamma_{desh}$, $Q = \gamma_{desh}(\gamma_{exc} + \gamma_{spon}) + \gamma_{non}(\gamma_{desh} + \gamma_{exc})$ and the constant

$$K = \frac{k_- + \gamma_{desh} - \gamma_{exc}\frac{\gamma_{non}}{\gamma_{desh}}}{k_+ - k_-}. \qquad (4.45)$$

The $g^{(2)}$ function described by Eq. 4.44 exhibits *anti-bunching* for $\tau = 0$, while *bunching* appears at medium timescales. At certain times after a photon detection, the detection of a consecutive photon is more likely. This can be explained, since for sure the system is not trapped in the metastable deshelving state after a photon detection. In contrast to the average, the occupation of the deshelving state does not vanish, resulting in less photon emission.

As a last remark, Eq. 4.44 depends only on three independent parameters and it is not possible to determine the complete dynamics from a measurement of the autocorrelation function.

4.2 Coherent Interactions

In the last section, light–matter interaction was treated in second-order perturbation theory, i.e., applying Fermi's golden rule. Within this approximation it was already possible to describe phenomena like absorption and emission correctly. Nevertheless, so far, quantum mechanic coherences, i.e., the off-diagonal elements of the density matrix, were neglected. This chapter will make up for this neglect, resulting in the concept of microscopic polarizations and the well-known Rabi oscillations.

4.2.1 *Optical Bloch Equations*

In this section the equations of motion for a single two-level system interacting with a strong laser field of frequency ω, i.e., an electric field given by Eq. 3.23 are derived. When the driving field is sufficiently strong, the back action of the particle can be neglected and the electric field at the particles position evolves as

$$E(t) = E_0 \cos \omega t = \frac{1}{2} E_0 \left(e^{i\omega t} + e^{-i\omega t} \right). \tag{4.46}$$

With this, the interaction Hamiltonian $\mathcal{H}_{int}$ in the dipole approximation gets the form

$$\mathcal{H}_{int} = -\hat{D} \cdot E_0 \cos \omega t. \tag{4.47}$$

For the particle a ground state $|g\rangle$ and excited state $|e\rangle$ is assumed (Fig. 4.10(a)), likewise in Section 4.1. The density matrix of this system is $\hat{\rho} = |\psi\rangle \langle \psi|$, with the matrix elements

$$\hat{\rho}_{gg} = |g\rangle \langle g| \quad , \quad \hat{\rho}_{ee} = |e\rangle \langle e|, \tag{4.48}$$

$$\hat{\rho}_{ge} = \hat{\rho}_{eg}^\dagger = |g\rangle \langle e|. \tag{4.49}$$

Furthermore, it is convenient to define the inversion operator

$$\hat{\sigma}_z = \hat{\rho}_{ee} - \hat{\rho}_{ee}, \tag{4.50}$$

from which the individual densities can be reconstructed using $Tr(\rho) = 1$. Together with the identifications

$$\hat{\sigma}_+ = \hat{\rho}_{eg}, \tag{4.51}$$

$$\hat{\sigma}_- = \hat{\rho}_{ge} \tag{4.52}$$

the inversion fulfills a Pauli spin algebra:

$$[\hat{\sigma}_+, \hat{\sigma}_-] = \hat{\sigma}_z, \tag{4.53}$$

$$[\hat{\sigma}_z, \hat{\sigma}_\pm] = \pm 2\hat{\sigma}_\pm. \tag{4.54}$$

Using that by parity considerations $\langle g|\hat{D}|g\rangle = \langle e|\hat{D}|e\rangle = 0$, and defining the real dipole matrix element $d = \langle g|\hat{D}|e\rangle$, the Hamiltonian can be rewritten as

$$\mathcal{H} = \frac{\hbar \omega_0}{2} \sigma_z - \hbar \Omega (\sigma_+ + \sigma_-) \cos \omega t, \tag{4.55}$$

where the energy origin is set to be half the way between ground and excited state and the *Rabi frequency* Ω is defined as

$$\Omega = \frac{d \cdot E_0}{\hbar}. \tag{4.56}$$

So far, spontaneous emission was neglected. When considering spontaneous emission from the upper level with the rate $1/T_1$, the dynamics can be described by the *Lindblad equation*

$$\frac{d}{dt}\hat{\rho} = \frac{i}{\hbar} \left[\hat{\mathcal{H}}, \hat{\rho} \right] + \mathcal{L}(\rho). \tag{4.57}$$

with the Lindblad form

$$\mathcal{L}(\rho) = -\frac{1}{2T_1} \left(\sigma_+\sigma_-\rho + \rho\sigma_+\sigma_- - 2\sigma_-\rho\sigma_+ \right). \tag{4.58}$$

To expand the equations of motion for the components of the density matrix, the commutator relations Eqs. 4.53 and 4.54 and the fact that the system can be excited only once, i.e., $\sigma_+\sigma_+ = \sigma_-\sigma_- = 0$ are useful. Now, terms oscillating with a frequency of the order 2ω are assumed to average out, i.e., the *rotating wave approximation* (RWA) is established. Furthermore, introducing the detuning $\Delta = \omega_{ge} - \omega$, a rotating frame $\sigma_-'(t) = \sigma_-(t)e^{i\Delta t}$, and $T_2 = 2T_1$ the equations of motion read

$$\dot{\sigma}_-'(t) = \left(i\Delta - \frac{1}{T_2} \right) \sigma_-'(t) + \frac{i\Omega}{2} \cdot \sigma_z(t), \tag{4.59}$$

$$\dot{\sigma}_z(t) = 2\Omega \cdot \text{Im}(\sigma_-') - \frac{1}{T_1} \left[\sigma_z(t) + 1 \right]. \tag{4.60}$$

Making use of $\rho_{ee} + \rho_{gg} = 1$ and the definitions of σ_z and $\sigma_\pm$, the *optical Bloch equations* [29] are obtained:

$$\dot{\rho}_{ge}'(t) = (i\Delta - 1/T_2)\rho_{ge}'(t) + \frac{i}{2}\Omega \left[\rho_{ee}(t) - \rho_{gg}(t) \right], \tag{4.61}$$

$$\dot{\rho}_{gg}(t) = -\frac{i}{2}\Omega\rho_{ge}' + c.c. + 1/T_1\rho_{ee}, \tag{4.62}$$

$$\dot{\rho}_{ee}(t) = \frac{i}{2}\Omega\rho_{ge}' + c.c. - 1/T_1\rho_{ee}. \tag{4.63}$$

In most systems, unavoidable external fields give rise to small fluctuations of the energy levels. This gives rise to *dephasing*, i.e., additional destruction of the off-diagonal elements with rate $1/T_2'$. Thus, in Eqs. 4.59–4.63, the coherence time $T_2 = 2T_1$ has to be replaced by

$$T_2^* = \left(\frac{1}{T_2} + \frac{1}{T_2'} \right)^{-1}, \tag{4.64}$$

and the lifetime of the excited state gives only an upper limit for the coherence time T_2^*.

Although an optical two-level system was assumed in the derivation of the Bloch equations, it can also be applied to spins in a magnetic field, as shown in the following.

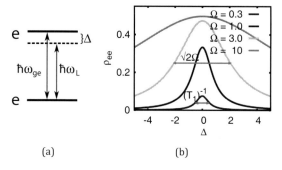

(a)   (b)

Figure 4.10 (a) Illustration of the considered levels and frequencies. Excited state e and ground state g are separated by $\hbar\omega_{ge}$, while the laser of frequency ω_l is detuned by Δ from the resonance condition. (b) Occupation of the excited level in the steady state, according to Eq. 4.67 as a function of detuning for $T_1 = 1$, $T_2^* = 2$. While for weak excitation the linewidth is $\Delta\omega = 2/T_2$, it approaches $\Delta\omega = \sqrt{2}\Omega$ under strong excitation.

4.2.2 Analogy to Spins in Magnetic Fields

Electron spins in magnetic fields are described by the Hamiltonian

$$\mathcal{H} = \frac{g\mu_B}{2}\sigma \cdot \mathbf{B}(t), \qquad (4.65)$$

with the *Bohr magneton* $\mu_B = e\hbar/(2m_e c) \approx 5.8 \cdot 10^{-5}$ eV/T and the *Landé g-factor* $g \approx 2$ for electrons. Assuming that the B_z component is static and without loss of generality $B_x = B_0 \cos(\omega t)$, while $B_y = 0$ this can be rewritten by using $\sigma_\pm = \sigma_x \pm i\sigma_y$ to

$$\mathcal{H} = \frac{g\mu_B}{2}B_z\sigma_z + \frac{g\mu_B}{4}B_0\cos(\omega t)(\sigma_+ + \sigma_-). \qquad (4.66)$$

In the considered case, the Hamiltonian is exactly of the form Eq. 4.55 and hence the Bloch equations can also be applied to spins in magnetic fields.

4.2.3 Steady-State Solution

In the steady state, the left hand side of Eqs. 4.61–4.63 vanishes. Furthermore, Eq. 4.63 can be eliminated by the conservation of the total probability, i.e., using $\rho_{gg}(t) + \rho_{ee}(t) = 1$. In this case, the excited state occupation is given by

$$\rho_{ee} = \frac{T_1\Omega^2}{2T_2}\frac{1}{(T_1\Omega^2 + 1)/T_2^* + \Delta^2}, \qquad (4.67)$$

i.e., a Lorenzian of width

$$\Delta\omega = \frac{2}{T_2^*}\sqrt{T_1 T_2^* \Omega^2 + 1}. \tag{4.68}$$

When increasing the excitation intensity $I \sim \Omega^2$ in resonance $(\Delta = 0)$, the excited state occupation shows a saturation behavior with a maximal value of $\rho_{ee} = 0.5$. This equals the result derived from first-order perturbation theory when assuming resonant pumping, i.e., using Eq. 4.17.

In the weak excitation limit $(1/T_2 \gg \Omega)$ the resonance is given by a Lorentzian of width $\Delta\omega = 2/T_2^*$ (Fig. 4.10). Hence, in the absence of dephasing the resonance width is given by $\Delta\omega = 1/T_1$, as expected from first-order theory Eq. 4.32.

Under strong excitation $(2/T_1 = 1/T_2^* \ll \Omega)$ the width of the resonance is $\Delta\omega = \sqrt{2}\Omega$. This phenomenon is widely known as *power broadening* [30, 31]. Here, stimulated emission effectively reduces the excited state lifetime and thereby increases the linewidth, as illustrated in Fig. 4.10(b).

4.2.4 *Rabi Oscillations*

When the Rabi frequency Ω dominates the dynamics, i.e., $\Delta \approx 1/T_2^* \approx 1/T_1 \approx 0$, Eq. 4.59 and Eq. 4.60 describe a harmonic oscillation between inversion and the imaginary part of the coherence. Coherences are converted into inversion and vice versa, as illustrated in Fig. 4.11(a).

In this case, with the initial condition $\sigma_z(0) = -1$ the inversion follows a cosine with Rabi frequency Ω. Hence, also the occupation of the individual levels oscillates, which is well known as *Rabi oscillations*. When applying the laser field for the time $t = \pi/\Omega$ the populations of the two levels are exactly swapped, i.e., the inversion flips the sign. Such a pulse is named π-pulse and is an important tool in many experiments on coherent manipulation of two-level systems.

Apart from this trivial case, two special cases, when either the detuning Δ or the dampings $1/T_1$, $1/T_2$ are negligible, are particular instructive. These cases are investigated in the following sections after introducing the Bloch sphere in the next section.

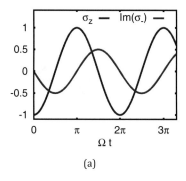
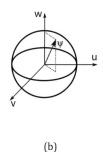

(a)   (b)

Figure 4.11 (a) Evolution of the inversion between excited and ground state $\sigma_z = \rho_{ee} - \rho_{gg}$ and the coherence $\rho_{ge} = \sigma_-$ with $\sigma_z(0) = -1$ under continuous coherent excitation. (b) The Bloch sphere, as a representation of the pure states of the two-level system. The projection on the w axis gives the inversion, $w = \sigma_z/2$, while the u and v axes represent, respectively, the real and imaginary part of the coherence $\rho_{eg} = \sigma_-$.

4.2.5 Bloch Vector

The *Bloch vector* **U** named after F. Bloch is a geometric representation of the density matrix defined by inversion σ_z and the coherence σ_-. It is defined by the three Cartesian components u, v, w with

$$u + iv = \sigma_-, \tag{4.69}$$

$$w = \sigma_z/2. \tag{4.70}$$

One can show that all pure states represented by a 2×2 density matrix ρ, i.e., that fulfill $Tr(\rho^2) = 1$, correspond to a Bloch vector of length $1/2$. The set of all pure states forms the Bloch sphere illustrated in Fig. 4.11(b). As this is applicable to any 2×2 density matrix, it hold in particular for all spin-1/2 systems.

The classical pure states, e.g., $\rho_{gg} = 1$ and $\rho_{ee} = 1$, correspond to the poles of the Bloch sphere. Starting from there, quasi-resonant Rabi oscillations are represented by rotations around the u axis, while the detuning Δ results in rotations around the w axis. This behavior is the basis of most magnet resonance and spin echo experiments. Nevertheless, in general the dynamics are complicated and eventually require a numerical solution of the Bloch equations. Details can be found in *nuclear magnetic resonance* (NMR) textbooks like Refs. [32–34].

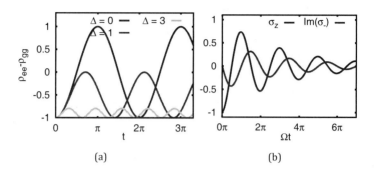

(a)   (b)

Figure 4.12 (a) Evolution of the inversion $\sigma_z = 2w$ with $2w_0 = \sigma_z(0) = -1$ with detuning $\Delta/\Omega = (0, 1, 3)$ (b) Damped Rabi oscillations with $T_2 = 5/\Omega$.

4.2.6 Undamped Rabi Oscillations with Detuning

In absence of damping and an external field, i.e., when $\Omega = 0$ the coherence ρ'_{eg} rotates in the complex plane with frequency Δ. This rotation is nothing else but the beating between the microscopic dipole represented by $\rho_{eg}(t)$, oscillating with ω_{ge} and the external field oscillating with ω_L.

When an external field is present, one finds for the solution of Eqs. 4.59 and 4.60:

$$\sigma_z(t) = \sigma_z(0) \left[\left(\frac{\Omega}{\Omega_R} - 1 \right) + \frac{\Omega}{\Omega_R} \cos(\Omega_R t) \right], \qquad (4.71)$$

with the initial condition $\sigma_-(0) = 0$ and the effective Rabi frequency

$$\Omega_R = \sqrt{\Delta^2 + \Omega^2}. \qquad (4.72)$$

Hence, the detuning effectively reduces the contrast of the oscillations, while the oscillation frequency is enhanced. This behavior is illustrated in Fig. 4.12(b).

4.2.7 Static Decoherence

Decoherence is the decay of the quantum mechanic coherences like ρ_{ge}, i.e., the loss of the well-defined phase relation between different states. In the previous Section 4.2.1 on the Bloch equations, this effect was phenomenologically introduced via the coherence time T_2^*. In most systems this coherence time is extremely short due to strong coupling to the macroscopic environment. Consequently, only a few well-isolated quantum systems allow the observation of coherent dynamics, while most systems appear classical. While the ultimate limit for T_2^* is given by the relaxation time T_1, the latter one can be neglected in most solid state systems and hence the decoherence by coupling to the environment is given by the constant T_2'.

In most cases the decoherence rate $1/T_2^*$ can be assumed to be time independent. When furthermore $1/T_2^*$ is much smaller than the Rabi frequency Ω and the detuning Δ can be neglected, this results in damped Rabi oscillations and the inversion follows

$$\sigma_z(t) = \sigma_z(0)e^{-t/(2T_2^*)}\cos(\Omega t). \tag{4.73}$$

This very common case, which can be observed in many experiments, is illustrated in Fig. 4.12(b).

4.2.8 *Measurement-Induced Decoherence*

While Eq. 4.73 holds for time-independent decoherence, it is possible to artificially introduce decoherence at well-defined times, e.g., via projective measurements. This case is experimentally considered in Section 7.7. When performing a measurement of the systems energy, this projects the wave function onto an energy eigenstate. Herein the probability of measuring a certain eigenvalue ϵ_i is given by

$$P(\epsilon_i) = Tr(\hat{\rho}\hat{P}_{\epsilon_i}) \tag{4.74}$$

with the definition of the *projector* onto state $|\epsilon\rangle_i$

$$\hat{P}_{\epsilon_i} = |\epsilon\rangle_i \langle\epsilon|_i = \hat{\rho}_{\epsilon_i}. \tag{4.75}$$

After the measurement the system is in the corresponding eigenstate $|\epsilon\rangle_i$ and all subsequent energy measurements will give the same eigenvalue ϵ_i. The new density matrix of this realization of the experiment after the measurement then is

$$\hat{\rho}' = \frac{\hat{P}_{\epsilon_i}\hat{\rho}\hat{P}_{\epsilon_i}}{P(\epsilon_i)}. \tag{4.76}$$

Repeating the experiment over and over again, the ensemble density matrix after the measurement can be constructed as

$$\hat{\rho}'' = \sum_i P(\epsilon_i) \frac{\hat{P}_{\epsilon_i} \hat{\rho} \hat{P}_{\epsilon_i}}{P(\epsilon_i)} = \sum_i P(\epsilon_i) \hat{P}_{\epsilon_i}. \qquad (4.77)$$

This $\hat{\rho}''$ is nothing, but a density matrix consisting of all diagonal elements of the original density matrix. Hence, a measurement simply results in the destruction of all off-diagonal elements. Aside from this interpretation based on the *collapse of the wave function*, i.e., the *Copenhagen interpretation* of quantum mechanics and the *von Neumann measurement* process, it can also be seen as decoherence by coupling to a classical reservoir. In the measurement process the quantum system gets strongly coupled to the classical measurement apparatus. This induces strong decoherence and thus all off-diagonal elements disappear.

4.2.9 *The Quantum Zeno Effect*

As discussed above, measurements lead to the collapse of the wave function and hence the disappearance of off-diagonal elements in the density matrix. This can heavily influence the systems dynamics. In particular, *permanent* observation might hinder unstable quantum system from decaying as proposed in in 1977 by Misra et al. [35]. This *quantum Zeno paradox* is in close analogy to the classical Zeno paradox formulated no later than 430 BC by the ancient Greek philosopher Zeno of Elea [36]. If an object can be observed at any instance of time to be at a well-defined position in space, motion cannot occur. It took over two millennia until I. Newton and G. W. Leibniz solved the paradox with the invention of infinitesimal calculus. For the quantum Zeno paradox it took only a few years until it was shown that the condition of *permanent* observation is unphysical due to time–energy uncertainty arguments [37]. Nevertheless, if measurements are performed continuously, coherent Rabi oscillations can be completely frozen [38]. Here the continuous observation requirement is not as strict as in the original proposal by Misra et al. This *quantum Zeno effect* was first observed by the group of the 2012 Nobel laureate D. J. Wineland, who inhibited microwave-induced quantum jumps between two $^9\text{Be}^+$ ground-state hyperfine levels by repeated measurements [39]. For example, if a measurement is performed in the middle of a π-pulse, when the inversion σ_z is zero and the coherence σ_- reaches its maximum value of $\sigma_- = \pm i/2$, further coherent evolution is inhibited and according to Eq. 4.71 the inversion remains zero after. Using this mechanism, a coherent population transfer with a π-pulse can be inhibited with 50% efficiency.

Nowadays this quantum Zeno effect is mostly interpreted as decoherence by coupling to the macroscopic environment [40–42]. Aside from philosophic implications of the phenomenon, recent proposals aim toward technologically utilizing the broader formulated quantum Zeno dynamics [43] for efficient quantum gates and fault-tolerant quantum operation [44–46]. Later on, in Section 7.7, this is further investigated.

4.3 Three-Level Systems

While the last sections were devoted to the coherent dynamics of two-level systems, three-level systems are treated in this section. Here, the equations of motion for the three-level system interacting with two laser fields are derived in Section 4.3.1. While a general analytic solution of the dynamics cannot be given, the important example of *stimulated Raman transitions* is given in the following Section 4.3.2.

4.3.1 *The Λ System*

In this section, the Λ-type three-level system interacting with two coherent light fields as illustrated in Fig. 4.13 is investigated. Similar to the previous section, it is assumed that the driving fields are sufficiently strong to neglect back action. Now, there are two fields $E_{1(2)}$, each evolving as

$$\mathbf{E}(t)_{1(2)} = \mathbf{E}_{1(2)} \cos \omega_{1(2)} t. \tag{4.78}$$

With this, the interaction Hamiltonian $\mathcal{H}_{int}$ in the dipole approximation gets the form

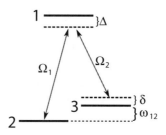

Figure 4.13 Sketch of a Λ-type three-level system, interacting with two electrical fields.

$$\mathcal{H}_{int} = -\frac{1}{2} \sum_{i \neq j} \hat{\mathbf{d}}_{ij} \cdot (\mathbf{E}_1 \cos \omega_1 t + \mathbf{E}_2 \cos \omega_2 t), \qquad (4.79)$$

where the sums run over the ground states $|2\rangle$ and $|3\rangle$, separated by $\hbar\omega_{23}$, and the excited state $|1\rangle$. The 3×3 density matrix of the system is given by $\hat{\rho} = |\psi\rangle \langle\psi|$, with the six independent matrix elements

$$\hat{\rho}_{11} = |1\rangle \langle 1|, \quad \hat{\rho}_{22} = |2\rangle \langle 2|, \quad \hat{\rho}_{33} = |3\rangle \langle 3|, \qquad (4.80)$$

$$\hat{\rho}_{12} = |1\rangle \langle 2|, \quad \hat{\rho}_{13} = |1\rangle \langle 3|, \quad \hat{\rho}_{23} = |2\rangle \langle 3|. \qquad (4.81)$$

When assuming that the transition between ground states does not couple to the fields, i.e., $\langle \hat{\mathbf{d}}_{23} \cdot \mathbf{E}_{1\,(2)} \rangle = 0$, and neglecting spontaneous emission as well as dephasing, the dynamics of the density matrix can be obtained from the Heisenberg equation, i.e., Eq. 4.57 with $\mathcal{L}(\rho) = 0$. Applying furthermore the RWA and transforming into a proper rotating frame, one finds

$$\dot{\rho}_{11} = -\dot{\rho}_{22} - \dot{\rho}_{33}, \qquad (4.82)$$

$$\dot{\rho}_{22} = -\frac{i}{2}\tilde{\rho}_{12}\left(\Omega_1 + \Omega_2\chi^*\right) + \text{c.c.}, \qquad (4.83)$$

$$\dot{\rho}_{33} = -\frac{i}{2}\tilde{\rho}_{13}\left(\Omega_1\chi + \Omega_2\right) + \text{c.c.}, \qquad (4.84)$$

$$\dot{\tilde{\rho}}_{12} = -\frac{i}{2}\left((\rho_{22} - \rho_{11})\left(\Omega_1 + \Omega_2\chi\right) + \tilde{\rho}_{32}\left(\Omega_1\chi^* + \Omega_2\right)\right)$$
$$-i\Delta\tilde{\rho}_{12}, \qquad (4.85)$$

$$\dot{\tilde{\rho}}_{13} = -\frac{i}{2}\left((\rho_{33} - \rho_{11})\left(\Omega_1\chi^* + \Omega_2\right) + \tilde{\rho}_{23}\left(\Omega_1 + \Omega_2\chi\right)\right)$$
$$-i\left(\Delta + \delta\right)\tilde{\rho}_{13}, \qquad (4.86)$$

$$\dot{\tilde{\rho}}_{23} = \frac{i}{2}\left(\tilde{\rho}_{21}\left(\Omega_1\chi^* + \Omega_2\right) - \tilde{\rho}_{13}\left(\Omega_1 + \Omega_2\chi^*\right)\right) - i\delta\tilde{\rho}_{23}. \qquad (4.87)$$

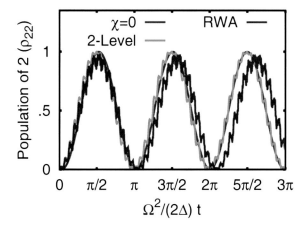

Figure 4.14 Numerical solution of the equations of motion, when the system is initially in state $|3\rangle$, i.e. $\rho_{33} = 1$ for $\Omega_1 = \Omega_2 = \Omega$, $\Delta = 6\Omega$, $\delta = 0$ and $\omega_{23} = 4/3\Omega$. When only the RWA is applied (Eqs. 4.82–4.87), the population oscillates between the ground states with a stepwise behavior (curve RWA). For the effective two-level system (Eqs. 4.90–4.92) the steps persist (2-level). Only when $\chi = 0$, i.e., each field couples only to one transition by RWA or selection rules, the oscillation gets sinoidal ($\chi = 0$).

Here, $\chi(t) = e^{i(\omega_{23}+\delta)t}$ when both allowed transitions couple equally to both fields, i.e., $\langle \hat{\mathbf{d}}_{12} \cdot \mathbf{E}_{1(2)} \rangle = \langle \hat{\mathbf{d}}_{13} \cdot \mathbf{E}_{1(2)} \rangle = \hbar\Omega_{1(2)}$, while $\chi = 0$ when each field couples only to one transition, e.g., by selection rules. So far, without any serious approximation these equations are quite general and with different initial conditions and assumptions on the frequencies, detunings and coupling strength lead to a plethora of phenomena like *coherent population trapping* (CPT), and *electromagnetic-induced transparency* (EIT). Furthermore, the equations of motion allow for *stimulated Raman adiabatic passage* (STIRAP) and *stimulated Raman transitions* (SRT) to coherently manipulate the ground state populations. This latter technique is introduced in the following.

4.3.2 *Stimulated Raman Transition*

When the detuning Δ is large compared to the Rabi frequencies $\Omega_{1(2)}$ and δ ($\Delta \gg \Omega_{1(2)}, \delta$), and the population of the excited state $|1\rangle$ vanishes initially, i.e., $\rho_{11}(0) = 0$, this state will never become populated. In this case the approximations $\rho_{11} = \dot{\rho}_{11} = \dot{\tilde{\rho}}_{12} = \dot{\tilde{\rho}}_{13} = 0$ holds and from Eqs. 4.85 and 4.86 one obtains

$$\tilde{\rho}_{12} \approx -\frac{1}{2\Delta}\left(\rho_{22}\left(\Omega_1 + \Omega_2\chi\right) + \tilde{\rho}_{32}\left(\Omega_1\chi^* + \Omega_2\right)\right), \quad (4.88)$$

$$\tilde{\rho}_{13} \approx -\frac{1}{2\Delta}\left(\rho_{33}\left(\Omega_1\chi^* + \Omega_2\right) + \tilde{\rho}_{23}\left(\Omega_1 + \Omega_2\chi\right)\right). \quad (4.89)$$

Placing this in Eqs. 4.83, 4.84 and 4.87 gives the dynamics of an effective two-level system:

$$\dot{\rho}_{22} = i\frac{\Theta}{2}\tilde{\rho}_{23} + \text{c.c.}, \quad (4.90)$$

$$\dot{\rho}_{33} = -\dot{\rho}_{22}, \quad (4.91)$$

$$\dot{\tilde{\rho}}_{23} = -i\frac{\Theta^*}{2}\left(\tilde{\rho}_{33} - \tilde{\rho}_{22}\right) - i\delta\tilde{\rho}_{23}, \quad (4.92)$$

where the effective Rabi frequency Θ is defined by

$$\Theta = \left(\left(\Omega_1^2 + \Omega_2^2\right)\chi + \Omega_1\Omega_2\left(1 + \chi^2\right)\right)/(2\Delta). \quad (4.93)$$

When either $\chi = 0$, or the influence of $\chi(t)$ averages out because the other dynamics are slow enough, the effective Rabi frequency gets time-independent. In this case, the dynamics get equivalent to the Bloch equations of a two-level system with the Rabi frequency $\Theta = \Omega_1\Omega_2/(2\Delta)$. This is known as stimulated Raman transition [47–51] and widely used in atomic physics. In this work, it is the basis of the entanglement scheme discussed later on in Section 15.2.

4.4 Cavity Quantum Electrodynamics

In the previous sections, free space was assumed. Hence, the electric field associated with the normal modes had the form Eq. 2.5 in the classical formulation, or Eq. 3.2 in the formalism of quantum mechanics. Although this is a proper description of empty space, mirrors or complex dielectric arrangements as described in Part III may mix normal modes with same frequency and different **k** vectors. In this case the free-space photon is not a good concept anymore, and it is better to transform to a new basis. This is done in the next Section 4.4.1, followed by a discussion of the physics arising from this in Sections 4.4.2 to 4.4.5.

4.4.1 Cavity Modes

As said above, dielectric arrangements can mix the normal modes and the operators are not orthogonal anymore. To construct new, re-diagonalized field operators $\hat{b}_m^\dagger$, a *Bogoljubov transformation* can be applied:

$$\hat{b}_m^\dagger = \sum_l U_m^l \hat{a}_l^\dagger, \qquad (4.94)$$

with a unitary norm preserving transformation U_m^l and l running over all modes with $\omega_l = \omega_m$. One can show [11] that the new operators $\hat{b}_m$, $\hat{b}_m^\dagger$ also fulfill the commutator relation Eq. 2.54, while the Hamiltonian is given by

$$\mathcal{H}_R' = \sum_l \hbar\omega_l \left[\hat{b}_l^\dagger \hat{b}_l + \frac{1}{2} \right]. \qquad (4.95)$$

Hence, the new operators give an equivalent description of the quantized electromagnetic field, and everything stated in Chapter 3 holds also for $\hat{b}_m$, $\hat{b}_m^\dagger$.

Electric Field of Cavity Modes The only difference between the free-space formulation and the cavity-mode formulation of quantum electrodynamics (QED) is the form of the electric field observable. While for $\hat{a}_m$ and $\hat{a}_m^\dagger$ the electric field was given by Eq. 3.2, now the electric field operator of a single mode m is

$$\hat{\mathbf{E}}_m(\mathbf{r}) = \mathcal{E}_m \mathbf{v}_m(\mathbf{r}) \hat{b}_m + \text{h.c.}, \qquad (4.96)$$

with the *normalized spatial mode profile*

$$\mathbf{v}_m(\mathbf{r}) = \sum_l i\epsilon_l U_m^l e^{i\mathbf{k}_l \cdot \mathbf{r}} \qquad (4.97)$$

describing the spatial distribution of the field. Here, the form of the unitary transformation guarantees that the mode profile is normalized according to Eq. 2.3, i.e.,

$$\int d\mathbf{r}\, |v_m(\mathbf{r})|^2 = V. \qquad (4.98)$$

Although it is in general not possible to find an analytic expression for U_m^l, the classical field modes $\mathbf{E}_m(\mathbf{r})$ can be found using techniques of computational electrodynamics, as done in Part III. By a proper normalization the classical modes give the normalized spatial mode profile

$$\mathbf{v}_m(\mathbf{r}) = \mathbf{E}_m(\mathbf{r}) \cdot \frac{V}{\int d\mathbf{r}\, \varepsilon(\mathbf{r})\, |\mathbf{E}_m(\mathbf{r})|^2}, \tag{4.99}$$

where $\varepsilon(\mathbf{r})$ denotes the dielectric constant at position $\mathbf{r}$, as introduced in Part III. Furthermore the *mode volume* V_{eff} as a measure of the mode confinement can be defined as

$$V_{eff} = \frac{\int d\mathbf{r}\, \varepsilon(\mathbf{r})\, |\mathbf{E}(\mathbf{r})|^2}{\max\left(\varepsilon(\mathbf{r})\, |\mathbf{E}(\mathbf{r})|^2\right)}. \tag{4.100}$$

While the volume of the free space modes equals the quantization volume V it can be as small as $\sim \lambda^3$ in nanostructures. With these definitions the electric field can be reformulated to

$$\hat{\mathbf{E}}_m(\mathbf{r}) = \mathcal{F}(\mathbf{r})\mathcal{E}'_m \hat{b}_m + \text{h.c.}, \tag{4.101}$$

with the electric field per photon given by

$$\mathcal{E}'_m = \sqrt{\frac{\hbar\omega_m}{2\varepsilon_0 V_{eff}}}, \tag{4.102}$$

and the form factor $\mathcal{F}(\mathbf{r})$ giving the polarization and ratio between field maximum and field at position $\mathbf{r}$

$$\mathcal{F}(\mathbf{r}) = \frac{\mathbf{E}_m(\mathbf{r})}{\max\left(\varepsilon(\mathbf{r})\, |\mathbf{E}_m(\mathbf{r})|^2\right)}. \tag{4.103}$$

Equation 4.102 is the key result of this section. The electric field generated by single photons or vacuum fluctuation in the mode maximum can be tremendously enhanced when the mode volume is small. In contrast, by choosing a structure such that $\mathbf{E}_m(\mathbf{r})$ is much smaller than its free space value, vacuum fluctuations can be suppressed.

One of the most important dielectric arrangements are *cavities*. These have the useful property that within a certain frequency range called *free-spectral range* there is only one single mode with frequency ω_c that generates an electric field inside the cavity, while all other modes are suppressed. Hence a two-level system resonant with the cavity, i.e., with energy splitting $\hbar\omega_c$, can couple only to this specific mode, while systems off-resonant to the cavity do not couple to the field modes at all.

In the following, the resonant case will be further investigated, yielding the key results of cavity quantum electrodynamics.

Cavity Damping   In section 4.1 it was assumed that emitted photons are immediately absorbed by a detector or the laboratory walls. Hence a back action onto the emitter was impossible. Although this is realistic for free space experiments, it is wrong for experiments with cavities. There the field is strongly confined in the cavity region, and photon losses occur only with the rate κ. Instead of using κ, cavities are usually described by the dimensionless *quality factor*

$$Q = \frac{\omega_c}{\kappa} = \frac{\omega_c}{\delta\omega_c}. \tag{4.104}$$

For the last transformation the time–energy uncertainty argumentation from Section 4.1.7 was used to identify the cavity damping κ with the spectral width of the cavity resonance $\delta\omega_c$.

4.4.2 Jaynes–Cummings Model

In section 4.2 an external coherent single mode photon field was assumed to drive a two-level system. There, the external field was assumed to be strong and hence back action of the two-level system onto the photon field was negligible. In this section this simplification is overcome by introducing the so-called *Jaynes–Cummings model*. As there is plenty of literature on the Janes–Cummings model [52–54] and cavity quantum electrodynamics, here the deviation is outlined only briefly.

Using the definitions of σ_z and $\sigma_\pm$ Eqs. 4.50–4.52 and defining the zero point of the energy to be halfway between excited and ground state, the particle Hamiltonian $\hat{\mathcal{H}}_P$ of the two-level system gets

$$\hat{\mathcal{H}}_P = \frac{1}{2}\hbar\omega_{ge}\hat{\sigma}_z. \tag{4.105}$$

Assuming that diagonal elements of the dipole operator vanish by parity arguments, e.g., $\langle g|\hat{d}|g\rangle = \langle e|\hat{d}|e\rangle = 0$ the interaction Hamiltonian gets

$$\hat{\mathcal{H}}_I = \hbar g_0 \left(\hat{\sigma}_+ + \hat{\sigma}_-\right)\left(\hat{b} + \hat{b}^\dagger\right), \tag{4.106}$$

with the *cavity coupling* constant

$$g_0 = \Omega_0 = \mathcal{E}'\epsilon \cdot \frac{\langle g|\hat{d}|e\rangle}{\hbar}, \tag{4.107}$$

which is also called *vacuum Rabi frequency*. Assuming $\omega \approx \omega_{ge}$ and using the Heisenberg equation Eq. 4.57 one can show that $\hat{\sigma}_+\hat{b}^\dagger$ and

$\hat{\sigma}_- \hat{b}$ rotate with 2ω and hence average out in the RWA. Together with the Hamiltonian of the radiation $\hat{\mathcal{H}}_P$, where the zero-point energy is dropped, one finds the *Jaynes–Cummings Hamiltonian*

$$\hat{\mathcal{H}}_{JC} = \frac{1}{2}\hbar\omega_{ge}\hat{\sigma}_z + \hbar\omega_c\hat{b}^\dagger\hat{b} + \hbar g_0\left(\hat{\sigma}_+\hat{b} + \hat{\sigma}_-\hat{b}^\dagger\right). \quad (4.108)$$

The following sections deal with the equations of motion arising from this Hamiltonian.

4.4.3 One-Photon Bloch Equations

Analogous to section 4.1.4, it is assumed that at maximum a single excitation is present and the two-level system in the cavity is hence fully described by the three joint states $|1\rangle = |e, 0\rangle$, $|2\rangle = |g, 1\rangle$, $|3\rangle = |g, 0\rangle$ and their superpositions. Considering cavity losses, the equation of motion for the density matrix is given by the Lindblad equation [13]

$$\frac{d}{dt}\hat{\rho} = \frac{i}{\hbar}\left[\hat{H}, \hat{\rho}\right] - \frac{\kappa}{2}\left(\hat{b}^\dagger\hat{b}\hat{\rho} + \hat{\rho}\hat{b}^\dagger\hat{b} - 2\hat{b}\hat{\rho}\hat{b}^\dagger\right). \quad (4.109)$$

Assuming a proper rotating frame for the polarization ρ_{12} and defining the cavity detuning $\Delta = \omega_{ge} - \omega_c$, this can be expanded to the *one-photon Bloch equations*:

$$\frac{d}{dt}\hat{\rho}_{11} = \frac{i}{2}\Omega_0\rho_{12} + \text{c.c.}, \quad (4.110)$$

$$\frac{d}{dt}\hat{\rho}_{22} = -\frac{i}{2}\Omega_0\rho_{12} + \text{c.c.} - \kappa\rho_{22}, \quad (4.111)$$

$$\frac{d}{dt}\hat{\rho}_{33} = -\kappa\rho_{22}, \quad (4.112)$$

$$\frac{d}{dt}\hat{\rho}_{12} = \frac{i}{2}\Omega_0\left(\rho_{11} - \rho_{22}\right) + \left(i\Delta - \frac{\kappa}{2}\right)\rho_{12}. \quad (4.113)$$

Using $\rho_{33}(t) = 1 - \rho_{11} - \rho_{22}$ the equations can be simplified, diagonalized, and solved. In the following sections, this is done for the case of negligible cavity damping ($\kappa = 0$), resulting in the phenomenon of *vacuum Rabi splitting*, and for the resonant case ($\Delta = 0$) giving rise to a discussion of *vacuum Rabi oscillations* and the *Purcell effect*.

4.4.4 *Vacuum Rabi Splitting*

By diagonalizing the equations of motion Eqs. 4.110, 4.111 and 4.113, the eigenstates and eigenfrequencies can be found with the ansatz function $e^{\lambda t}$. When photon losses from the cavity can be neglected ($\kappa = 0$), the *dressed* eigenstates of the coupled two-level cavity system represent good quasi-particles, the so-called *polaritons*. The imaginary part of the two non-zero eigenvalues $\lambda_{\pm}$ give the frequency shift of the polaritons with respect to the frame rotating with ω_{ge}:

$$\Im(\lambda_{\pm}) = \frac{1}{2}\left(\Delta \pm \sqrt{\Delta^2 + \Omega_0^2}\right). \quad (4.114)$$

These frequency shifts are related to an energy shift $\Delta E = \hbar\lambda_{\pm}$ plotted in Fig. 4.15(b). With this, the energy eigenvalues of the polaritons are found to be

$$E_{\pm} = \hbar(\omega_{ge} + \Im(\lambda_{\pm})). \quad (4.115)$$

In the uncoupled case, when $\Delta \gg \Omega_0$ the eigen energy E_- corresponds to the level splitting of the two-level system $\hbar\omega_{ge}$, while E_+ is the cavity photon energy $\hbar\omega_c$. In contrast, when coupled, the energy splitting between the two eigenstates equals twice the vacuum Rabi frequency $\Delta E = 2\hbar\Omega_0$ in the resonant case, i.e., for $\Delta = 0$. This behavior is known as vacuum Rabi splitting.

4.4.5 *Vacuum Rabi Oscillations and Purcell Effect*

Under the assumption that $\rho_{11}(0) = 1$ and hence all other matrix elements vanish at $t = 0$, the dynamics of ρ_{11} is given by

$$\rho_{11}(t) = e^{-\frac{\kappa t}{2}}\left[\cosh\left(\frac{At}{2}\right) + \frac{\kappa}{A}\sinh\left(\frac{At}{2}\right) - \frac{2\Omega_0^2}{A^2}\right], \quad (4.116)$$

with $A = \sqrt{\kappa^2 - 4\Omega_0^2}$. Depending on the ratio between κ and Ω_0, the strong coupling and the weak coupling regime can be distinguished.

Strong Coupling ($\kappa \ll \Omega_0/2$) In the strong coupling regime $\kappa \ll \Omega_0/2$, i.e., the coherent dynamics dominate over the incoherent losses. In this case $A \approx -2i\Omega_0$ and $\kappa/\Omega_0 \approx 0$ holds and Eq. 4.116 simplifies to

$$\rho_{11}(t) = \frac{1}{2}e^{-\kappa t/2}\left[1 + \cos\left(\Omega_0 t\right)\right]. \quad (4.117)$$

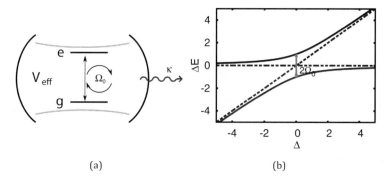

(a)   (b)

Figure 4.15 (a) Model system of cavity quantum electrodynamics. The electric field is confined to the cavity volume V_{eff}, resulting in the coupling Ω_0 between excited state e and ground state g. Cavity photons may be lost with rate κ to the environment. (b) The polariton energies Eq. 4.115 when varying the detuning Δ for $\Omega_0 = 1$ (solid lines) and $\Omega_0 = 0$ (dashed lines). For zero detuning, the coupling prevents the crossing of the two polaritions.

The upper-level population shows a damped oscillation. The two-level system goes to the ground state while it emits a photon into the cavity mode. Subsequently, this photon is absorbed again, and so forth. Hence the population oscillates between ρ_{11} and ρ_{22}, until the photon is finally lost from the cavity mode, as illustrated in Fig. 4.16. This is commonly known as vacuum Rabi oscillations and can also be interpreted as a beating of the two polariton modes, having slightly different frequencies.

Weak Coupling ($\kappa \gg \Omega_0/2$) When the damping is strong compared to the cavity coupling, the system is in the *weak coupling* or *Purcell regime*. In this case, for all terms linear in A the approximations $A \approx \kappa$ and $\Omega_0/\kappa \approx 0$ hold. In contrast, in the arguments of the *cosh* and *sinh* functions this approximation is too rough. There, the first-order Taylor expansion of A around $\Omega_0 = 0$, i.e., $A \approx \kappa - 2\Omega_0^2/\kappa$ must be used instead and hence:

$$\rho_{11}(t) = e^{-\Omega_0^2/\kappa t}. \tag{4.118}$$

No oscillation occurs and the occupation of the excited state is damped with the rate

$$\gamma_{Purcell} = \Omega_0^2/\kappa. \tag{4.119}$$

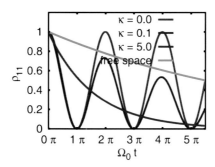

Figure 4.16 Vacuum Rabi Oscillations and Purcell Effect. The excited state population ρ_{11} is plotted as a function of time for different damping rates κ, given in units of Ω_0. Without damping ($\kappa = 0$), the population oscillates between ρ_{11} and ρ_{22}. This can be understood as a beating of the two polariton states. With damping, populations get lost to ρ_{33}, resulting in damped oscillations according to Eq. 4.117. For $\kappa > \Omega_0$ the excited state population is damped faster than in free space, according to Eq. 4.119. No oscillation occurs.

Comparing this to the spontaneous emission rate in free space Eq. 4.23 gives the Purcell factor [55]:

$$F = \frac{\gamma_{Purcell}}{\gamma_{spon}} = \mathcal{F}(\mathbf{r}) \cdot \epsilon_D \frac{3\lambda^3}{4\pi^2} \frac{Q}{V_{eff}}, \tag{4.120}$$

where λ is the wavelength in the cavity and ϵ_D the unit vector in direction of the dipole moment.

When the quantum system is well aligned with the cavity mode, the form factor $\mathcal{F}$ gets unity and strong enhancements of the spontaneous emission can be reached in small mode volume cavities with high Q.

PART II

QUANTUM SYSTEMS FOR INTEGRATION INTO HYBRID DEVICES

Introduction

In recent decades plenty of optical active quantum systems have been intensely studied on the level of single emitters. Integration into hybrid systems was demonstrated for single atoms, semiconductor quantum dots, color centers in wide-bandgap material, and single molecules. Among these, single atoms and ions in the gas phase played a pioneering role, and they were used to establish many experimental techniques. Already in the late 1990s the groups around S. Haroche in Paris and H. Walther in Garching coupled single atoms in the gas phase to ultra-high-finesse microwave cavities. In their experiments several basic concepts of cavity quantum electrodynamics, like vacuum Rabi oscillation [56, 57], were tested, and the EPR-like entanglement of two atoms was achieved [58]. Later on, the group around G. Rempe in Munich succeeded in building hybrids of optical cavities and single rubidium atoms. Here, cavity-enhanced cooling and trapping was demonstrated [59, 60].

Today, experiments with single atoms and ions are quite advanced, but there are only a few ideas and demonstrations of how they can be extended toward a scalable and integrable platform for quantum technology applications [61, 62]. The main reason is the demand for complex trapping and cooling setups including numerous lasers and ultra-high vacuum apparatuses, making the experiments hard to integrate. Thus, atomic hybrid systems are not in the scope of this work. In contrast, the main advantage of solid state systems is that trapping mechanisms are in general not required. Therefore, they are almost ideal for building integrable hybrid devices. Nevertheless, in most cases this is dearly traded against much poorer coherence properties.

In the following chapters, quantum dots, single molecules, and color centers in diamond are reviewed with special attention on their usage in experiments and applications. Recently, in particular, the color centers emerged as a promising resource for integrated quantum technology. Hence, the discussion goes into more detail here and numerous experimental results on nitrogen vacancy centers in diamond are presented. Section 7.3 treats experiments on fundamental optical properties, like single-photon emission, excited state lifetime, and the optical spectrum. The latter one is treated in more detail in Section 7.4 on spectral diffusion. Then, in Section 7.5 the fundamental spin physics of the NV center is introduced, before the impact of strain is discussed in Section 7.6. Part II concludes in Section 7.7 with an experimental demonstration of the quantum Zeno effect on a single NV center.

Chapter 5

Quantum Dots

Quantum dots (QDs) are small quasi-zero-dimensional regions of semiconductor in which the carriers are confined in a volume comparable to their de Broglie wavelength [63].

Such regions can be formed by combining two semiconductors materials, 1 and 2, with different bandgaps E_{g1} and E_{g2}. In such heterostructures, the relative positions of the conduction and valence bands are determined by the *electron affinities*, i.e., the energetical distance between conduction band and vacuum $\chi = E_c - E_{vac}$. If the lower conduction-band edge and the higher-valence band edge are both in the material with the smaller bandgap, the materials form a so-called type I heterostructure (Fig. 5.1).
In this case, the band discontinuities for the conduction and valence band are given by

$$\Delta E_c = \chi_1 - \chi_2, \tag{5.1}$$

$$\Delta E_v = (\chi_1 + E_{g1}) - (\chi_2 + E_{g2}). \tag{5.2}$$

These discontinuities for several typical material combinations are given in Table 5.1.

A very common technique to prepare quasi-zero-dimensional type I heterostructures is *self-assembling* [65, 66]: When a flat layer of a semiconductor is grown on top of a flat substrate (bulk) with

Integrated Quantum Hybrid Systems
Janik Wolters
Copyright © 2015 Pan Stanford Publishing Pte. Ltd.
ISBN 978-981-4463-82-9 (Hardcover), 978-981-4463-83-6 (eBook)
www.panstanford.com

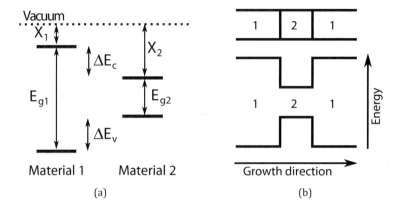

Figure 5.1 (a) Position of the band edges in a type I heterostructure. (b) Schematic of an overgrown quantum dot structure.

Table 5.1 Heterostructure materials

Material combination	ΔE_c/meV	ΔE_v/meV
$Al_{0.25}Ga_{0.75}As$/GaAs	311	36
GaAs/$In_{0.5}Ga_{0.5}As$	420	220
$In_{0.5}Ga_{0.5}As$/InAs	410	18.5

Band discontinuities at $T = 300$ K for several material combinations that form type I heterostructures. The material first noted has the larger bandgap. All values are taken from Ref. [64].

a slightly different lattice constant, the resulting strain may induce the formation of small ($\lesssim$ 60 nm) islands. If this so-called Stranski–Krastanov growth mode is reached, quite homogeneous quantum dot ensembles can be grown [67].

5.1 Quantum Dot Wavefunction and Level Structure

In what follows, at first the single-particle Hamiltonians of quantum dots in the envelope approximation [68] is examined. Because of the three-dimensional potential well, bound carrier states are formed within a quantum dot. In the spirit of the *envelope function approximation* the potentials of the individual atoms can be omitted and, for further simplification, the confinement potential can be approximated by an infinite potential. Thus the carrier in the quantum dot resembles the particle-in-a-box problem. A more accurate possibility is the use of a finite confinement potential or the assumption of a harmonic potential. The corresponding wave functions can be found in Ref. [19]. There are also several approaches using combinations of these two potentials [69]. The calculation of more realistic wave functions is a topic of its own, where details can be found in Refs. [70, 71].

For the simple *particle-in-a-box* approach, in one dimension the potential is given by

$$V_{con}(x) = \begin{cases} 0 & : |x - x_0| \leq L/2 \\ \infty & : |x - x_0| > L/2 \end{cases}.$$ (5.3)

The solutions of the Schrödinger equation with the potential given by Eq. 5.3 are easily calculated. Outside the quantum dot the wave function $\tilde{\chi}(x)_n$ vanishes, while one finds

$$\tilde{\chi}(x) = \begin{cases} \sqrt{2/L}\cos(k_n(x - x_0)) & n \text{ odd} \\ \sqrt{2/L}\sin(k_n(x - x_0)) & n \text{ even} \end{cases},$$ (5.4)

inside the quantum dot. For the eigenenergies one finds

$$\varepsilon_n^\lambda = \hbar v_k^\lambda = \frac{\hbar^2 k_n^2}{2m_\lambda} \text{ with } k_n = \frac{n\pi}{L}.$$ (5.5)

Combined with the Bloch functions $u(x)$ from the atomic structure of the semiconductor, the wave function of the bound QD states reads $\chi(x) = \tilde{\chi}(x)u(x)$.

Although this simple model is only a very rough approximation, it gives a good qualitative description of the resulting energy level structure: In the quantum dot several discrete localized levels exist. According to Pauli's principle, each can be occupied by two carriers of opposite spin. To get a more accurate description of the level structure, the Coulomb interaction must be included into the model. The electrostatic attraction of electrons and holes will allow for bound states, the so-called *excitons* (X). Furthermore, Coulomb repulsion changes the energy of globally charged states, i.e., states with different electron and hole number. Thus, the Coulomb interaction leads to a more complex structure of exciton, charged exciton, biexciton, etc. [72]. To observe these states in

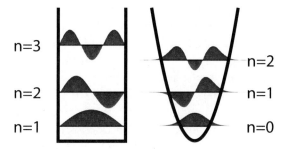

n=3

n=2

n=1

n=2

n=1

n=0

Figure 5.2 Schematic energy levels and the first three wave functions in an infinite potential well and an infinite harmonic potential as simple models for electron confinement in QDs. For the finite potential well, only few bound states exist.

experiments it is mostly necessary to keep the thermal energy $k_B T$ well below the binding energy of the quantum dot states. For Stranski–Krastanov grown quantum dots in III–V material, this requires cryogenic environments as the semiconductor continuum is energetically close to the QD states.

Chemically synthesized, so-called colloidal quantum dots offer the advantage of a deeper trapping potential and thus show photoluminescence even at room temperature. However, due to a poorly defined or controlled surface they show photobleaching and blinking [66]. Therefore such colloidal QDs are less used for integrated quantum technology application and are not treated here.

Another special type are electrically defined QDs. Here the confinement potential is dynamically generated using electrodes on a semiconductor surface [73]. While this allows the use of almost any semiconductor material, the electrodes limit the integrability into more complex hybrid devices.

5.2 Experiments with Single Quantum Dots

The most important feature common to all quantum dots is the relatively large dipole moment of the optical exciton recombination. The exciton lifetime is usually on the order of 1 ns, making QDs important candidates as gain media in lasers and other optoelectronic devices. Apart from this classical applications, QDs are an important resource for optical quantum technology. The three major applications are QD single photon sources, entangled photon sources, and spin qubits, which are reviewed in the following.

5.2.1 *Single Photon Source*

Today, the main application of single photons is *quantum key distribution* (QKD) [74]. Here, messages are secured by using the laws of quantum mechanics. This is regarded to be much safer than the classical encryption algorithms, which mainly rely on unsolved mathematical problems, i.e., factoring numbers [75]. In order to be unconditional secure, in most QKD schemes a one-time encryption key is distributed using single photons [76].

In general, such single photons can be generated by excitation and subsequent fluorescence of individual quantum systems. Here, care has to be taken that really a *single* excitation in a *single* system is created. If optical excitation is used, this is guaranteed by using a confocal configuration [77] and spectral filtering. If the QD concentration is low enough, only a single QD is in the confocal volume. The excitation laser generates electron–hole pairs in the vicinity of the QD. By Coulomb and phonon scattering the carriers will relax into the QD ground state, where an exciton is formed. This exciton recombines under emission of a single photon. Alternatively, when several excitons are present, the multi-exciton emission can be filtered spectrally.

While in colloidal QDs this process is possible at room temperature [5], epitactic QDs usually require a cryogenic environment because of the weaker confinement potential [79]. Apart from this disadvantage, epitactic QDs are usually preferred, since they can be easily integrated into p-i-n diode structures for electrical pumping. Such electrical pumping is clearly preferable, since no external pump sources are necessary. Furthermore, most applications like quantum key distribution require single photons on demand and the electrical triggering is easier achievable than optical triggering. For such electrically pumped QD single photon sources, there are two main challenges: The current path must be controlled in order to pump only one single quantum dot and the dot must be filled equally with electrons and holes to prevent emission from charged

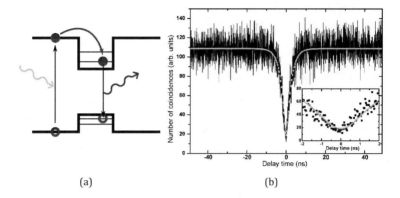

(a)   (b)

Figure 5.3 (a) Schematic of the process leading to the emission of a single photon: A laser generates electron–hole pairs. One of the pairs is scattered into the QD ground state, where it forms an exciton. This exciton recombines under emission of a photon. Eventually, photons from multi-exciton decays can be filtered out spectrally. (b) Measured autocorrelation of the photons emitted from the exciton line of a single InP QD. The inset shows a magnification of the region around zero delay. The dip clearly shows the absence of multi-photon events. From [78], © 2003 APS.

excitons. The first is usually tackled by placing a current aperture atop the quantum dot, while for the second one, the position of the QD within the pin structure is crucial [80, 81]. Today, these problems are frequently solved and currently much research effort is put into the development of ultra-bright and spectrally very narrow sources by using of hybrid structures [81, 82]. A further important task is the development of quantum dots emitted at telecommunication wavelength, as required for QKD applications.

Table 5.2 Optical parameters of typical quantum dots

material	wavelength/nm	exciton lifetime/ns	Ref.
InAs / GaAs	850–1000	∼ 1	[83]
CdSe / ZnSe	500–550	∼ 0.2	[83]
InP / InGaP	650–750	∼ 1	[83]

Emission wavelength and excited state lifetimes of typical quantum dots

5.2.2 Entangled Photon Source

Another important application of single QDs is their potential to generate *entangled photon pairs*.

To generate such pairs the *biexciton* (XX) formed by two electron–hole pairs inside the QD is used. If such a biexciton is generated, it will decay in a cascade process via the single exciton an thereby subsequently emits two photons. To understand why these two photons can be entangled, it is important to have a closer look on the carrier spins in the QD: The two electrons have spin $J_z = \pm\frac{1}{2}$, while the heavy holes have spin $J_z = \pm\frac{3}{2}$. Now, the electrons recombine with the holes of opposite spin under emission of a circular polarized photon. Thereby in the biexciton cascade the anticorrelated electron spins can be translated into anticorrelated polarizations of the photons [84]. Nevertheless, the photons are only entangled if the fine structure splitting between the different polarized excitons vanishes. In this case photons from the left and right path in Fig. 5.4a have equal energy and the resulting photon state is

$$|\Psi^+\rangle = \frac{1}{\sqrt{2}}\left(|\sigma^+\rangle_1\,|\sigma^-\rangle_2 + |\sigma^-\rangle_1\,|\sigma^+\rangle_2\right).$$

This is a maximally entangled *Bell state*.

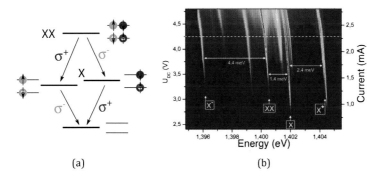

(a)   (b)

Figure 5.4 (a) If the quantum dot is charged with two electron–hole pairs, a biexciton (XX) is formed. The biexciton decays via the single exciton (X), either under emission of left (σ^+) or right circular (σ^-) polarized light. (b) Electroluminescence of a single quantum dot. At higher injection currents above 1 mA the biexciton peak and charged excitons gets visible. Fine structure splitting is not resolved. With kind permission from Ref. [384].

In most quantum dots, the two polarizations have a slightly different transition energy due to fine structure splitting. This results in a quantum beating of the entangled state. To avoid this beating it is important to control the fine structure splitting of the exciton. While the first experiments using InAs/GaAs QDs relied on chance during the growth process [85, 86], today it seems that this problem can be deterministically solved by simultaneous control of mechanical strain via the piezoelectric effect and electrostatical tuning via the Starck-effect [87] in suitable nanostructures.

5.2.3 *Spin Qubit*

A third possible application is the use of the electron spin of a *charged quantum dot* as spin qubit for quantum computation [88]. For this, several criteria must be fulfilled:

(1) The QD must provide a spin.
(2) The spin coherence time T_2 must be sufficiently long to perform many coherent operations.
(3) It must be possible to initialize the QD into a definite spin state.
(4) Some form of spin read out must be provided.

The first requirement (1) is always met in charged quantum dots, where the additional electron provides the necessary spin 1/2. Here, a suitable donor, e.g., silicon, is implanted in the vicinity of the QD layer. The donor provides excess electrons, which charge some of the QDs with a single electron [89]. Requirement (2) is problematic for QDs made of III–V semiconductors, where the nuclear spins inevitable introduce an inhomogeneous magnetic field. Even at low temperatures this limits the T_2^* usually to the order of 10 ns, while spin diffusion limits the coherence time T_2 to about 10 μs [90]. Here the spin-free group V semiconductors (Si, Ge) are clearly preferable, where QD can be electrically defined by proper gate electrodes [91, 92].

To initialize the charged QD into a definite spin state, the negatively charged exciton, i.e., the *trion* can be used [93]. In detail, a resonant circular polarized π-pulse is used to pump the state $|\uparrow_e\rangle$ into the state $X^{-\downarrow} = |\downarrow_e \uparrow_e \Downarrow_h\rangle$, where the electron and hole will recombine subsequently and leave the QD in the state $|\downarrow_e\rangle$

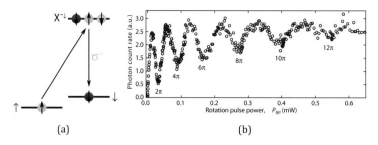

(a)   (b)

Figure 5.5   (a) Scheme of the initialization of a charge QD into state $|\downarrow\rangle$: If QD is initially in the state $|\uparrow\rangle$, it is pumped by a resonant π-pulse into the trion state $|X^{-\downarrow}\rangle$ and subsequently decays into the desired state $|\downarrow\rangle$. By proper choice of the polarization and frequency, pumping of the undesired trion $|X^{-\uparrow}\rangle$ is prevented. (b) Optically induced Rabi oscillation of spin of a single electron in a InGaAs QD. The final state of the QD oscillates with the square rot of the rotation pulse power. From Ref. [89], © 2008 Nature Macmillan Publishers Ltd.

(Fig. 5.5(a)). Using external magnetic fields on the order of a few hundred mT to a few T, this state can be Zeeman-splitted from its counterpart $|\uparrow_e\rangle$ by about 1 GHz. With resonant magnetic [94] or optical fields in the Raman configuration [89, 93] it is possible to coherently manipulate the spin.

Subsequent readout is possible by selectively driving the transition to the trion and collect the emitted single photons from the $|\uparrow_e\downarrow_e\Uparrow\rangle_h \Rightarrow |\uparrow_e\rangle$ transition. Here, the single shot readout fidelity is limited by the photon detection efficiency which can be boosted by the use of optical cavities. Alternatively, using the spin blockade mechanism and an ancilla QD in the vicinity, the spin can be read out electrically [94].

Chapter 6

Single Molecules

Single organic molecules have several important properties, that make them excellent quantum emitters. Compared to other emitters, they usually show bright fluorescence with high quantum efficiencies. Furthermore, the emission lines can be spectrally very narrow and the emission of different molecules covers a broad spectral range.

Nevertheless, the optical observance of single organic molecules in the condensed phase was long time an unsolved task. This had several reasons. First, the inevitable substrates and matrices produces fluorescence and Raman scattering background light, which easily dominates the overall fluorescence. Second, during the excitation and fluorescence process molecules undergo transformations, leading to their distraction. If irreversible, this process is known as *photobleaching*, while reversible processes lead to *blinking* of the molecule. Only by the late 1980s and early 1990s these problems could be solved by the availability of suitable laser sources and efficient detection schemes [4, 95]. Before describing experiments with single molecules in Section 6.2, the fundamentals of single molecule physics are provided in the following Section 6.1.

Integrated Quantum Hybrid Systems
Janik Wolters
Copyright © 2015 Pan Stanford Publishing Pte. Ltd.
ISBN 978-981-4463-82-9 (Hardcover), 978-981-4463-83-6 (eBook)
www.panstanford.com

6.1 Fundamentals of Single Molecules

Optically active organic molecules are built of the antisymmetric *p-wave functions* of carbon atoms. These electronic orbitals are largely extended and easily bond to other p orbitals, forming the so-called π bonds. Using this mechanism the large variety of organic molecules is constructed [Fig. 6.1(a)].

Within the molecules the electrons can be seen to move almost freely, resulting in a low-dimensional potential well, comparable to the particle in the box problem of section 5.1. If light is absorbed, the electronic cloud is put into the excited state. In this excited state usually the equilibrium position of the nuclei is slightly different, leading to the excitation of vibrational modes after optical excitation or de-excitation of the electrons. This is known as the *Franck–Condon principle* and gives rise to vibronic sidebands in fluorescence, as well as absorption [Fig. 6.1(b)]. The purely electronic transitions, where no vibronic modes participate are called *zero phonon line* (ZPL). While at room temperature, these vibronic transitions are usually broadened, leading to an absorption and fluorescence continuum, at low temperatures sharp line can be observed. In case of similar potentials in the ground and excited

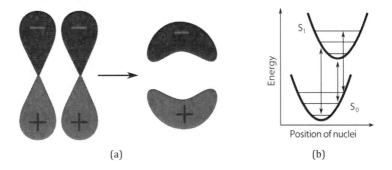

(a) (b)

Figure 6.1 (a) The angular part of the antisymmetric p orbitals projected into the plane. Two p orbitals can form a so-called π bond. Plus and minus refer to the sign of the wave function, rather than charge. (b) Illustration of the Franck-Condon principle. The ground state S_0 and excited state S_1 exhibit different equilibrium positions of the nuclear position. This leads to participation of the vibronic sublevels in the optical transition between the states.

state the vibronic bands in absorption and fluorescence are mirror symmetric around the ZPL. Ideally absorption and fluorescence are conjugated processes, and the emission and absorption rates are equal. Nevertheless, in most cases the excited state has several possible decay channels, leading to larger decay rate. If these additional channels are nonradiative, the *quantum yield* η can be significantly reduced. In order to be a good quantum emitter the nonradiative channels must be suppressed, as it is the case in most fluorescing dyes. Furthermore, it must be noticed, that the π-electrons are easily polarized, giving high chemical activity. In the excited state this is even worse, resulting in high instability of the molecule. After emission about 10^6 photons this leads usually to the death of the molecule [96]. To prevent this, the molecule can be cooled down, embedded into a solid matrix [97] or investigated under oxygen free atmosphere [96]. Another issue is the existence of *metastable states*. If the molecule reaches a metastable state during the excitation cycles, the fluorescence will drop suddenly and recovers only after some time.

In order to perform systematic studies on single molecules, or to integrate them into hybrid structures, they must be immobilized, i.e., be in the condensed phase. This can be done for example by *spin-coating* the dissolved molecules and matrix material onto a low

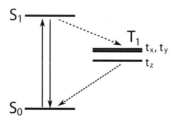

Figure 6.2 The electronic level structure of pentacene, a typical optical active molecule. The molecule has a singlet ground state S_0 and a singlet excited state S_1. Between these states optical transition with high oscillator strength take place, as indicated by the solid arrows. In addition, a triplet state T_1 with the sub levels $t_{x,y,z}$ exists. This state can be reached via radiation-free inter system crossing (ISC), indicated by dashed arrows. The rates for these ISC processes are much lower than for the radiative transitions.

fluorescence substrate, like fused silica or clean glass. Here typically films of a few ten nm thickness are achieved with a spinning speed of 1000–5000 rpm. In general, the orientation of the molecule's dipole with respect to the surface cannot be controlled resulting in all possible orientations.

6.2 Experiments with Single Molecules

Today many groups all over the world investigate single molecules. In most of these experiments, the molecules are used as bright and narrow band single photon emitters. In the following Section 6.2.1, these properties are reviewed on the example of DBT embedded in an anthracene matrix, before spin properties of molecules are regarded later on in Section 6.2.2.

6.2.1 *Room Temperature Single Photon Source*

Due to the photo instabilities of organic molecules the research on single molecules concentrated for a long time on molecules bound in solids at liquid helium temperatures and below [98]. For example, in Ref. [99] a single dibenzanthanthrene (DBATT) molecule embedded into a hexadecane matrix is reported to show single photon emission for at least a few minutes at temperatures of 1.7 K. Even today, only a few optical active molecules that are more or less photostable at room temperature are known. The most important ones are terrylene in a p-terphenyl matrix [100] and *dibenzoterrylene* (DBT) in anthracence [97].

Table 6.1 Molecules Emitting Single Photons

molecule and host	λ	τ	T	Ref.
Terrylene in p-terphenyl	579 nm	~4 ns	300 K	[100]
DBT in anthracence	780 nm	~5 ns	300 K	[97]
DBATT in hexadecane	590 nm	~8 ns	1.7 K	[99]

Emission wavelength λ, excited state lifetimes τ and operating temperature T of hosted molecules exhibiting stable single photon emission.

In particular, the latter one is interesting as DBT can be incorporated into 20 nm thick films, inherently compatible with hybrid devices. For this, a solution of crystalline anthracene (AC) and DBT is spin coated onto a clean cover slide. In this films the AC matrix efficiently shields the DBT against oxygen and other quenchers, making them stable for about 4 hours, even under optical excitation. Single-photon emission from single DBT molecules can be excited by a 725 nm laser and detected in a confocal configuration. With an oil immersion objective of NA = 1.4 the single-photon count rate can reach 10^6 counts/s at saturation, making DBT one of the brightest reported single-photon sources. At room temperature the DBT molecules' fluorescence spectrum exhibits a peak at about 790 nm with 50 nm FWHM. In contrast, at temperatures below

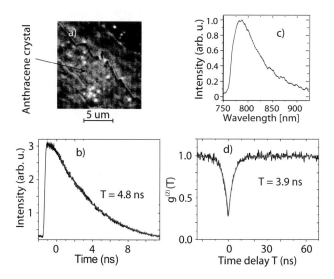

Figure 6.3 (a) Wide-field image of a sample of DBT in anthracene. The bright spots correspond to individual DBT molecules. (b) Fluorescence lifetime measurement of a single the DBT molecule. The excited state lifetime is 4.8 ns. (c) Room temperature fluorescence spectrum of a single DBT molecule. The fluorescence is centered around 790 nm, the position of the ZPL. (d) The measured second-order autocorrelation of the photons emitted by a single DBT molecule. The anti-bunching dip at zero time delay proves the single photon character of the emitted radiation. All adapted from Ref. [97]. © 2010 OSA.

2 K, the absorption line widths can be as narrow as 30 MHz, close to the lifetime limit [101–103]. This makes DBT promising for future experiments like two-photon interference or spectroscopy with single photons.

6.2.2 *Optically Detected Magnetic Resonance*

For application in quantum information processing spin states are of particular interest, since they provide longer coherence times, than optical transitions. Unfortunately, the optical active singlet ground states $S_{0,1}$ do not provide a spin degree of freedom. Only the triplet state T_0 can be used for spin experiments. Here the problem is that the triplet state of most molecules does not provide suitable optical transitions [104]. Nevertheless, it is possible to perform *optically detected magnetic resonance* (ODMR) measurements. The prerequisite for this is to optically detect fluorescence from a single molecule. During the emission and excitation cycles the molecule will occasionally get into the metastable triplet state, where it stays until the comparably slow ISC brings it back into the singlet ground state. As a consequence, the molecule emits photons in bunches, separated by dark periods in which the molecule is in the effectively dark triplet state. If the number of photons detected during one bright period is large enough, the ISC into the triplet can be directly observed as a *quantum jump* [105]. Otherwise, the dwell time in the triplet lowers the overall fluorescence intensity. Now, the different spin sublevels of the triplet state occur to have different population and depopulation rates. By applying a microwave field which is resonant with the fine structure splitting of the triplet, the population of the sublevels can be altered, leading to change of the average dwell time in the triplet. As a consequence, the average fluorescence intensity of the molecule is changed. A typical molecule to perform such experiments is pentacene in p-terphenyl host crystal cooled to 1.2 K. In this molecule, the z-polarization state is long lived and negligibly populated during the normal excitation cycles. Furthermore, at zero magnetic field it is split by approximately 1.4 GHz from the x and y-state [106, 107]. A microwave of this frequency can be applied via a short circuit loop at the end of a suitable coaxial cable and fluorescence intensity is

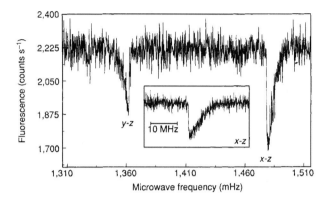

Figure 6.4 Fluorescence intensity of a single pentacene molecule when a microwave field is swept over the z–x and z–y resonance of the triplet state. The inset shows a magnification of the z–x resonance at reduced microwave intensity. Taken from Ref. [106]. © 1993 Nature Macmillan Publishers Ltd.

recorded as a function of the microwave frequency in the region around the magnetic resonance. If the microwave hits one of the resonances between x, y, and z state, the latter one gets populated, leading to an decreased fluorescence intensity.

With a similar technique called *electronuclear double resonance* (ENDOR) it is possible to address the nuclear spin states via their hyperfine coupling to the electronic resonances [108].

Chapter 7

Color Centers in Diamond

For quantum applications, diamond is one of the most remarkable materials. On one hand, its wide band gap of 5.5 eV makes it optically transparent from far infrared to deep ultraviolet. On the other hand, the face centered cubic lattice can host plenty of optical active impurities, the so-called *color centers*. The most prominent of these color centers is the *nitrogen-vacancy* (NV) center, giving many diamonds a yellowish color. As nitrogen is the most abundant impurity, it is somehow natural to classify diamond according to its nitrogen content. The four types of diamond that are commonly distinguished are listed in Table 7.1.

Table 7.1 Different types of diamond

Type	Nitrogen content	Comment
I a	≤ 1000 ppm	clusters of 3–5 N atoms
I b	≤ 500 ppm	single N atoms
II a	~ 0 ppm	ultrapure
II b	~ 0 ppm	boron impurities

Diamond is classified into four types, according to the nitrogen content.

Integrated Quantum Hybrid Systems
Janik Wolters
Copyright © 2015 Pan Stanford Publishing Pte. Ltd.
ISBN 978-981-4463-82-9 (Hardcover), 978-981-4463-83-6 (eBook)
www.panstanford.com

Apart from the nitrogen-vacancy center, today over 100 optical active impurities are known [28]. A few of them have been studied on the level of single centers, showing the capability of emitting single photons. Table 7.2 lists some of the most prominent ones. A more extensive review on diamond-based single-photon emitters can be found in Ref. [109].

Table 7.2 Single defects in diamond

Defect center	λ	τ	T	Ref.
nitrogen-vacancy (NV$^-$)	638 nm	~12 ns	300 K	[6, 110]
silicon-vacancy (SiV)	740 nm	~1.2 ns	300 K	[111, 112]
Ni-N complex (NE8)	793 nm	~2 ns	300 K	[113, 114]

Zero phonon emission wavelength λ, excited state lifetimes τ and operating temperature T of several color centers in diamond that have been studied on the single center level.

Another remarkable property of diamond rises from the fact that the ^{12}C carbon atoms in the diamond lattice, which have a natural abundance of 98.9% are intrinsically spin free. This makes diamond ideal for spintronic applications. Here, in particular centers providing triplet ground states, which can be initialized and read out optically, are in the research focus. Namely, the negatively charged NV center allows coherent manipulation of electronic spin states with coherence times in the millisecond regime. Thus it is not surprising that in the last decade the negatively charged NV center in diamond has emerged as a promising resource for future quantum technology [28, 90, 115–118]. Various experiments to coherently control NV spin qubits by microwave pulses have been performed. Notably, simple quantum gates have been demonstrated in cryogenic environment and even at room temperature [119]. It has also been possible to entangle the electronic spin with neighboring ^{13}C nuclear spins [120], with an emitted photon [121], or other NV spins [122, 123] over a distance of up to three meters. While in most experiments the NV's electron spin is coherently controlled by microwaves, all-optical spin manipulation is also feasible [121, 124–127].

7.1 Nanodiamond

Diamond is not only available in form of bulk crystals, but also in form of nanoscopic diamond monocrystals with a size of only a few ten nanometres, which are ideal to build quantum hybrid devices in a bottom up approach [128–132], as will be exploited in Part IV.

Such diamond nanocrystals (here referred to as nanodiamonds) can be synthesized on an industrial scale using high-pressure and high-temperature (HPHT) or detonation chamber growth. After synthesis and removal of contaminations and non-diamond carbon, the diamond particles feature an oriented crystal structure with parallel-running cleavage planes, similar to natural diamond. Such industrial diamonds are a popular choice for grinding, lapping, and polishing purposes. Furthermore, high-energy ball milling to produce nanodiamonds from bulk diamonds and several other synthesis methods are frequently used as well [133–135].

In principle, the optical properties of color centres in nanodiamonds are identical to those in bulk diamond. Likewise in bulk diamond, the NV center is the most abundant color center in nanodiamonds. An estimated 5% of typical commercially available nanodiamonds with a size of about 30 nm contain a single centre. By spin-coating of a properly diluted and ultrasonic treated diamond solution, individual nanodiamonds can be deposited on suitable substrates like cover slips cleaned with oxygen plasma, acid, or alkaline liquid. Eventually, small amounts of a polymer such as *polyvinyl alcohol* (PVA) can be added to improve distribution and sticking of the nanodiamonds to the substrate. Other centres, such as the silicon-vacancy (SiV) center, and nickel/chromium-related centers can be grown by chemical vapor deposition (CVD) using commercial nanodiamonds as seeds. Here, impurity atoms from the substrate are incorporated into the diamond nanocrystal during the growth process [111, 136, 137] or can be implanted by ion beams [138–142]. This results in an increased abundance of impurity atoms and the corresponding defect centers.

In the following two color centers which are observable in nanodiamonds are introduced. The next section 7.2 will briefly treat the *silicon-vacancy center* (SiV), before considering the nitrogen-vacancy center in detail from section 7.3 on.

7.2 Silicon-Vacancy Center in Diamond

The silicon-vacancy center consists if a single silicon impurity with two adjacent vacancies [143]. The crystallographic structure is illustrated in Fig. 7.1(a). These defects usually occur in CVD grown diamond, where silicon from the substrate or the quartz parts of the reactor is easily incorporated into the diamond lattice [144, 145]. Alternatively, silicon can be added intentionally or implanted into pure diamond by ion implantation [112]. Recently, the growth of SiV containing nanodiamonds of about 120 nm size from commercial diamond seeds on an iridium template could also be demonstrated [111].

Using confocal microscopy, individual SiVs can be identified by their single photon fluorescence at about 740 nm [cf. Fig. 7.1(b)] under excitation with a 671 nm laser focused tightly through a high NA objective lens. The *Debye–Waller factor*, i.e., the percentage of the overall emission going into the ZPL is on the order of 0.8, indicating

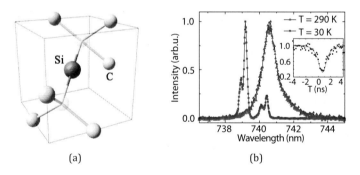

(a) (b)

Figure 7.1 (a) The crystallographic structure of the silicon-vacancy (SiV) color center in diamond. The SiV consists of a substitutional silicon atom with two adjacent vacancies. Taken from Ref. [109]. (b) The fluorescence spectrum of a single SiV at room temperature and 30 K. The inset shows the measured second order autocorrelation function $g^{(2)}(t, t+\tau)$ of the emitted photons, proving that a single center was measured. Taken from Ref. [111]. Both © 2011 IOP.

that most photons are actually emitted into the ZPLs. In saturation the count rates are on the order of 1000 counts/s for SiVs in bulk diamond [112], while SiVs in nano diamond are reported to be ultra bright with up to 5 million counted single photons per seconds. With this, SiVs in nanodiamond on iridium substrates are among the brightest photo-stable single photon sources available [111].

While at room temperature only one rather broad peak is visible in the spectrum, low-temperature measurements unravel details on the level structure of the SiV [111, 136, 145]. The existence of four lines separated by a few GHz can be attributed to a ground and excited state doublet. A detailed analysis of the population dynamics gives evidence for an even more complicated level structure consisting of four levels [137]: The optical active doublets, labeled 1 and 2, a metastable state 3, and a deshelving state 4 (cf. Fig. 7.2). Furthermore, the quantum efficiency η could be estimated to about 10%, i.e., in average only one photon is generated per 10 excitations of the SiV. This is a major drawback for using the SiV as on-demand single-photon source, but might be improved by the enhancement of the radiative transition via suitable nanostructures.

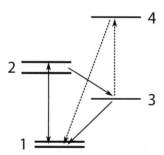

Figure 7.2 Level structure of the SiV. Spectroscopic measurement at low temperatures indicate the existence of two optical active doublets 1 and 2 being responsible for the fluorescence around 740 nm. The zero field splitting of the ground state 1 is on the order of 50 GHz, while the splitting of the excited state 2 is on the order of 240 GHz [145]. Since all transitions are possible, four lines can be observed in the spectrum. A detailed analysis of the population dynamics hints to a metastable state 3 with an excitation dependent deshelving mechanism via the state 4 [137].

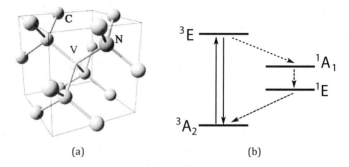

(a)   (b)

Figure 7.3 The nitrogen-vacancy center. (a) The crystallographic structure of the nitrogen-vacancy (NV) color center in diamond. The NV consists of an substitutional nitrogen atom with an adjacent vacancy. Taken from Ref. [109]. © 2011 IOP. (b) Simplified electronic-level structure of the NV center, without considering the spin fine structure. Nomenclature according to the NVs C_{3v} symmetry [149].

7.3 Nitrogen-Vacancy Center in Diamond

The nitrogen-vacancy center consists of a nitrogen impurity atom and an adjacent vacancy [Fig. 7.3(a)]. It belongs to the C_{3v} symmetry group, characterized by a threefold axis (C_3) with three vertical planes of symmetry (v). Naturally, it is the most abundant defect in diamond, and among all color centers it was the first one studied on the single center level in bulk diamond [6], as well as in nanodiamonds [110]. The NV center occurs in a neutral state (NV^0) and a charged state (NV^-). Usually particular interest is in the NV^- center, as the additional electron gives rise to a spin structure suitable for coherent quantum manipulation. The charge state conversion might be photo induced and plenty of research effort is spend to understand and control the conversion process [146–148]. Anyhow, in the following only the negatively charged NV^- center is treated, which will be referred to simply as NV center.

7.3.1 *Observation of Single Nitrogen-Vacancy Centers*

Single NV centers can be excited by a tightly focused 532 nm beam ($\sim$300 µW) to emit fluorescence between 625 and 775 nm. Using a confocal setup as illustrated in Fig. 7.4, this fluorescence can be

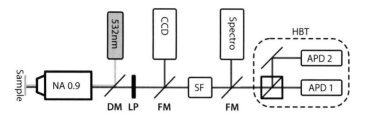

Figure 7.4 Optical setup for measurements on NV centers. Through an objective lens with high numerical aperture (NA 0.9) the sample is illuminated by a strong laser exciting the emitter (e.g., 532 nm, 300 μW). Fluorescence light is collected through the same objective lens and separated by a dichroic mirror (DM). Residual laser light is filtered by a long-pass filter (LP), while stray light is removed by a spatial filter (SF) consisting of a pinhole in a telescope prior detection by the spectrometer (Spectro) or a Hanbury Brown and Twiss interferometer (HBT). For adjustment, a CCD camera put into the optical path with a flip mirror (FM) is useful. To find single NV centers either sample or beam scanning must be implemented (not shown for simplicity).

detected and analyzed to gain insight into the emitter properties. By varying the excitation power, and monitoring the detected photon count rate, the saturation properties can be measured [Fig. 7.5(a)]. Here, the measured nonlinear behavior according to Eq. 4.27 is a clear witness of the quantum character of the studied system. This can be further supported by measuring the second order intensity autocorrelation function with a Hanbury Brown and Twiss setup, as discussed in Section 4.1.3. Alternatively photon number resolving detectors [24], or detectors with negligible dead time might be used [26]. Figure 7.5(b) shows a typical autocorrelation function measured on a single NV center. The *anti-bunching* dip at $\tau = 0$ indicates a strongly reduced probability for the detection of two photons simultaneously. This corresponds to a suppression of Fock states with $n > 1$, and hence proofs that the detected light predominantly stems from a single emitter. The drastically increased probability of detecting two photons with a time delay 20 ns $\lesssim$ $\tau \lesssim 1000$ ns, the so-called *bunching* peak is a clear signature of the multilevel scheme of the NV center illustrated in Fig. 7.3(b), as discussed in Section 4.1.9.

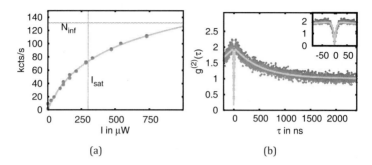

(a)   (b)

Figure 7.5 NV saturation behavior and autocorrelation function. (a) Saturation measurement on a bulk diamond NV center with solid immersion lens [150] according to Eq. 4.27. The fitting parameters are $I_{sat} = 299\ \mu W$, $\eta N_{inf} = 132$ kcts/s, $\eta a_{bg} = 25$ cts/μJ. (b) Measured $g^{(2)}$-function of a NV center. The fit according to Eq. 4.44 (solid line) gives $K = 0.98$, $k_+ = -0.18$/ns, $k_- = -0.0019$/ns. The inset show a zoom into the region around $\tau = 0$.

7.3.2 Excited State Lifetime and Spectral Properties

When using a pulsed laser instead of continuous wave excitation, it is possible to directly measure the excited state lifetime. Here the photon arrival time relative to the laser pulse gives the time dependent population probability of the excited state [Fig. 7.6(a)]. For the NV center in bulk diamond one usually finds lifetimes of about 11 ns, while in nanodiamonds the lifetime is on the order of 20 ns. This difference can be explained by the increased optical density of states in the high index bulk diamond, as discussed later on in Section 8.3.

Using a spectrometer with sensitive (e.g., liquid nitrogen cooled) detector, the spectrum of single NV centers can be resolved [Fig. 7.6(b)]. Apart from the slightly visible zero phonon line (ZPL) the spectrum appears almost featureless at room temperature. In contrast at temperatures of about 5 K, where phonon scattering is significantly reduced the ZPL gets prominent and its contribution to the overall intensity, i.e., the Debye–Waller factor is up to 5%.

Under this off-resonant excitation condition the typical line width of the ZPL in type Ib diamond is on the order of $\delta\lambda = 0.5$ nm, corresponding to $\delta f \approx 200$ GHz, far away from the Fourier limit of $\delta f \approx 10$ MHz given by the frequency uncertainty Eq. 4.30. In

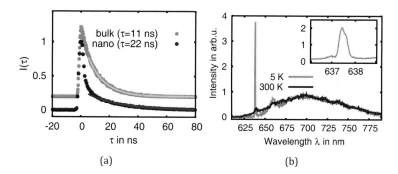

Figure 7.6 NV excited state lifetime and spectrum. (a) Measurement of the NV excited state lifetime in bulk and nanodiamond. The fit according to Eq. 4.28 gives $1/\gamma_{bulk} = 11$ ns, while in nanodiamond $1/\gamma_{nano} = 22$ ns due to the lower density of states. Owing to short lifetime background fluorescence, only data for $\tau > 5$ ns were used for the fit. (b) The fluorescence spectrum of a single NV in a nanodiamond at room temperature and 5 K. The inset shows a magnification of the low temperature ZPL.

contrast, in ultrapure type IIa bulk diamond almost lifetime limited line widths could be observed, allowing for two-photon interference experiments [151, 152]. Nevertheless, using resonant excitation with 637 nm line widths down to 16 MHz are reported even for type Ib nanodiamonds [153], although the lines were only short-term stable and appeared broader on a longer timescale.

The origin of the line broadening in type Ib diamond is spectral diffusion. Because of its importance for future experiments, it is extensively treated in the following section, based on Ref. [154].

7.4 Spectral Diffusion

Even in high-quality type IIa diamond, for implanted NVs located close to the surface, several GHz wide spectral diffusion can be observed under off-resonant illumination [155]. In nitrogen-rich type Ib nanodiamonds the energy levels fluctuate even stronger, as discussed earlier. It is widely assumed that such line broadening of emitters in condensed phase is due to a fluctuating electrostatic environment [156, 157].

In diamond nanocrystals these fluctuations are caused by ionizing impurities [158, 159] and charge traps. The spectral shift caused by single charges can be estimated using the simple toy model sketched in Fig. 7.7(a). Using data from Refs. [160, 161] one finds that the Stark shift of the excited states $\Delta\mathcal{E}_{x/y} = d \cdot E$ in an NV center induced by a single elementary charge in a distance of about 10 nm is as large as several hundred GHz and hence can explain the inhomogeneous line width.

The time after which the probability that the electrostatic environment remains unchanged is reduced to $1/e$ can be defined as the spectral diffusion time τ_D. Within this time τ_D the system can be assumed to be free from spectral diffusion, and all photons emitted within τ_D should be nearly indistinguishable. Enhancing this number of indistinguishable subsequent photons is a key goal for future applications.

Following Ref. [154] results on τ_D in NV centers in nanodiamonds under off-resonant excitation are presented in this section. As key result, these measurements identify the excitation laser as the main source for enhanced spectral diffusion.

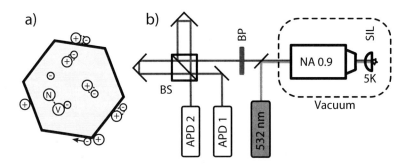

Figure 7.7 (a) Toy model for spectral diffusion. Charge traps are ionized and lead to fluctuating electric fields at the NV center's location. (b) Sketch of the experimental setup. Milled type Ib nanodiamonds are deposited on a solid immersion lens (SIL) and placed in a continuos-flow helium cryostat. Single-photon fluorescence is excited with a 532 nm laser and collected through an objective lens with numerical aperture NA = 0.9. Using a band-pass filter (BP), photons from the zero-phonon transition are filtered and sent through a folded Mach–Zehnder interferometer. From Ref. [154], © 2013 APS.

7.4.1 Techniques for Measuring Spectral Diffusion

The most straightforward method to measure spectral diffusion in single emitters is to take a time series of spectra and directly visualize the spectral wandering [157, 162, 163]. However, this method is only suitable for low spectral diffusion rates. For many emitters, like NV centers in nanodiamonds the rate of detected photons emitted from the ZPL transition is only on the order of a few kHz, while spectral diffusion occurs much faster.

In the literature three methods which in principle enable spectral diffusion measurements on timescales as fast as the emitter's lifetime were suggested and demonstrated recently [154, 164, 165]. Common to these techniques is the idea to convert the frequency fluctuations due to spectral diffusion into intensity fluctuations. However, in the first approach [164], technical limitations such as the required complex setup with several moving parts reduced the time resolution in the first measurements to the order of 100 μs [165]. In a related approach Sallen et al. [163] reached the sub-nanosecond regime, but for the price of a spectrometer-limited spectral resolution and an inherently reduced detection efficiency. As pointed out in Ref. [166] this approach also requires the emission frequency to fluctuate around one fixed central wavelength, which is in general not the case for quantum emitters.

In contrast, the recently developed *photon correlation interferometry* technique [154] uses an advanced and simplified interferometric setup, combining the advantages of the previous approaches. The timing resolution is limited only by the single photon counting instrumentation, and thus can be on the order of 100 ps. The spectral resolution can almost be as high as for conventional Fourier spectroscopy, while the central wavelength of the emission does not need to be fixed. Furthermore, the photon detection efficiency is at least doubled compared to the scheme presented in Ref. [163]. Therefore, this method is applicable even for low-intensity emitters with moderate integration times.

7.4.2 The Theory of Photon Correlation Interferometry

In photon correlation interferometry, a fixed *Mach–Zehnder interferometer* is used as disperse element, converting the spectral modulation of the ZPL into an intensity modulation. This intensity modulation can be measured with high accuracy by correlating the photons from the two interferometer outputs. Figure 7.7(b) shows a sketch of the setup. From the measured photon statistics, the timescale of the spectral diffusion is derived.

To understand how this works, the cross-correlation between the outputs of the interferometer under the influence of spectral diffusion is calculated in the following. Here, the mode index l of the photon number operator $\hat{n}_l(t)$ is suppressed for simplicity.

The time-averaged photon detection rate in the left or right output port of the interferometer is $< \hat{I}(t)_{L/R} >_t$, with

$$\hat{I}(t)_{L/R} = \eta_{L/R} \, \hat{n}(t) \, m_{L/R}(t), \tag{7.1}$$

where $\eta_{L/R}$ is the overall quantum efficiency in the left and right exit of the interferometer, respectively. $m_{L/R}(t)$ is the interferometer introduced modulation

$$m_{L/R}(t) = 1 \pm c \, \sin[2\pi x/\lambda(t)], \tag{7.2}$$

where c is the contrast of the interference fringes and x the path length difference of the interferometer arms. The plus and minus signs correspond to the left and right arm, respectively. The cross-correlation between the two arms $g_{LR}^{(2)}(\tau) = \frac{<:\hat{I}_L(t)\hat{I}_R(t+\tau):>_t}{<\hat{I}_L(t)>_t<\hat{I}_R(t)>_t}$ reads

$$g_{LR}^{(2)}(\tau) = g^{(2)}(\tau) < m_L(t)m_R(t+\tau) >_t, \tag{7.3}$$

where $g^{(2)}(\tau) = \frac{<:\hat{n}(t)\hat{n}(t+\tau):>_t}{<\hat{n}(t)>_t^2}$ is the second order autocorrelation function of the bare emitter and $<: \ldots :>_t$ denotes normal ordering before time averaging.

It is assumed that spectral diffusion occurs in form of jumps of the narrow emission line to random positions within a broad envelope [153], that the probability of a jump is independent of the spectral position, and that individual jumps lead to a significant change of the interferometer transmission. To calculate $< m_L(t)m_R(t + \tau) >_t$ the characteristic function $X(t, \tau)$, which is 1 if no spectral jump occurred within the time interval $(t, t + \tau)$ and 0 else is introduced and terms with and without spectral jump are separated before calculating the average. By using that $\lambda(t) = \lambda(t + \tau)$ if no jump occurred and $\lambda(t + \tau) = \lambda(t') \neq \lambda(t)$ if a jump occurred, separating $X(t, \tau)$ and identifying the probability that the spectral position remains unchanged after the time τ as $p(\tau) = < X(t, \tau) >_t$ one gets

$$< m_L(t)m_R(t + \tau) >_t = p(\tau) < m_L(t)m_R(t) >_t$$
$$+ [1 - p(\tau)] \cdot < m_L(t)m_R(t') >_{t,t'} . \quad (7.4)$$

If the interferometer is adjusted in a way that several fringes are within the inhomogeneous width of the ZPL, the sine terms average out and $< m_L(t) \cdot m_R(t') >_{t,t'} = 1$, while $< m_L(t)m_R(t) >_t = 1 - c^2/2$. Therefore,

$$< m_L(t)m_R(t + \tau) >_t = 1 - c^2/2 \cdot p(\tau). \quad (7.5)$$

Remarkably, here also path length fluctuations on timescales slower than the spectral diffusion average out and interferometric stability is not required. Finally, the probability that the spectral position of the ZPL remains unchanged after some time τ is

$$p(\tau) = 2/c^2 \cdot \left(1 - \frac{g_{LR}^{(2)}(\tau)}{g^{(2)}(\tau)} \right). \quad (7.6)$$

This equation indicated that it is sufficient to measure the autocorrelation function $g^{(2)}(\tau)$ of the bare emitter and the cross-correlation function between the two outputs L/R of the interferometer $g_{LR}^{(2)}(\tau)$ to gain knowledge on the spectral dynamics of the emitter.

7.4.3 Measurement of Spectral Diffusion by Photon Correlation Interferometry

To measure spectral diffusion, nanocrystals milled from high-quality type Ib bulk diamond with a size of 30 nm to 100 nm are spin coated on a *solid immersion lens* (SIL) (cf. Section 9.2) made of zirconium dioxide [167]. The SIL is placed in a continuous-flow He cryostat at about 5 K. Fluorescence is excited and collected in a confocal configuration [see Fig. 7.7(b)] through a commercial NA 0.9 objective lens placed inside the isolation vacuum of the cryostat. The NVs are excited by a green 532 nm laser with a power of several µW in front of the vacuum chamber. The single photons emitted by the NV are collected through the same objective lens and separated by a dichroic mirror from the excitation beam. An additional longpass filter blocks the excitation laser. In order to suppress the phonon side band, a removable bandpass filter (width 7 nm) centered at 637 nm is used. The filtered photons are directed into a Mach-Zehnder interferometer and detected by two avalanche photo diodes in the two outputs and counted with a time-correlated single photon counting module. The arm length difference of the interferometer is adjusted to obtain four fringes per nm at 639 nm, the ZPL wavelength (see Fig. 7.8). To check the alignment of the interferometer one of the APDs can be replaced by a 500 mm spectrograph with a cooled CCD detector and spectra are recorded (see Fig. 7.8).

In order to derive the probability $p(\tau)$ that the spectral position remains unchanged after the time τ according to Eq. 7.6 first the bare second-order autocorrelation of the NV emission has to be measured. In order to do this, the whole fluorescence from the NV including phonon side-bands is sent into the interferometer. In this case the autocorrelation function $g^{(2)}(\tau)$ measured by the two APDs is unaffected by the interferometer as can be proven by independent

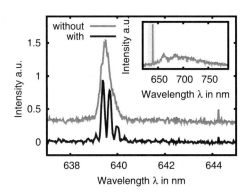

Figure 7.8 Spectra of the ZPL from a typical nanodiamond at a temperature of 5 K with a removable bandpass filter centered at the ZPL (637 nm). The upper curve (upshifted by 0.3 for clarity) is the NV center's fluorescence measured without the interferometer. The lower curve is recorded after the interferometer. The inset shows the full NV spectrum with the shaded area indicating the transmission of the bandpass filter. From Ref. [154], © 2013 APS.

measurements. The setup acts as a usual Hanbury Brown and Twiss setup. Next, the 637 nm bandpass filter is added in front of the interferometer and the cross-correlation function $g_{LR}^{(2)}(\tau)$ between the interferometer arms is measured with a typical integration time of 30 minutes. In this case the interferometer converts the spectral jumps of the narrow ZPL emission into intensity fluctuations. Care has to be taken not to change the laser power or adjustment within the total measurement time.

7.4.4 Results of Spectral Diffusion Measurements

Examples of measured $g^{(2)}$ and $g_{LR}^{(2)}$ functions taken at an excitation laser power of 14.1 µW are shown in Fig. 7.9(a). In order to derive the spectral diffusion probability $1 - p(\tau)$ the data is evaluated according to Eq. 7.6, resulting in the data shown in Fig. 7.9(b). Measurements taken with different interferometer adjustments resolving spectral jump widths from 20 GHz to 260 GHz showed, that the assumption of a narrow line jumping in a broad envelope is well justified (cf. Fig 7.10). Obviously, the chosen integration

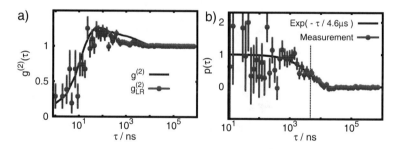

Figure 7.9 (a) Measured second-order autocorrelation function $g^{(2)}(\tau)$ (solid line) and cross-correlation function $g_{LR}^{(2)}(\tau)$ (dots) at an excitation laser power of 14.1 µW. The bin size is increased for longer delays τ, to decrease the error. The error the $g^{(2)}$ function is on the order of the line width. The dip at zero time delay shows the single-photon character of the emission. (b) The probability $p(\tau)$ that the ZPL remains constant on the order of 100 GHz within the time interval τ calculated from the dataset shown in (a) according to Eq. 7.6. The bin size is increased for longer delays τ, to decrease the error. The solid line is an exponential fit to the data showing a spectral diffusion time τ_D of (4.6 ± 0.6) µs, indicated by the dashed vertical line. From Ref. [154], © 2013 APS.

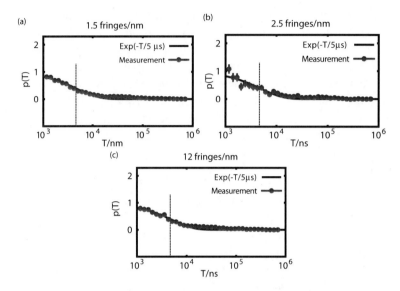

Figure 7.10 Spectral diffusion time at different interferometer positions. In (a) only spectral jumps larger than 260 GHz are resolved. In (b) the resolution is on the order of 110 GHz, while in (c) jumps larger than 20 GHz are resolved. The spectral diffusion time for all measurements is the same, indicating that spectral diffusion occurs in form of random jumps of the ZPL with a width below 20 GHz to a new position within the full broadened ZPL. From Ref. [154], © 2013 APS.

time of 30 min is sufficiently long to resolve the spectral diffusion, but if higher timing resolution is needed the integration time must be increased. The decrease of the probability $p(\tau)$ fits well to a single exponential decay. This supports the assumption that spectral diffusion is indeed caused by uncorrelated charge fluctuations. In the dataset shown in Fig. 7.9 the spectral diffusion time τ_D obtained from the fit is 4.6 ± 0.6 μs corresponding to a spectral diffusion rate $\gamma_D = 1/\tau_D$ of about 220 kHz, while detection rate at the APDs of photons from the ZPL transition is only 2.3 kHz.

The achieved contrast c derived from the fit is 35%, which is lower than the measured interferometer contrast of 90%. This is due to fluorescence background and the two nondegenerate NV dipole transitions $E_{x,y}$ [160] discussed in the following Section 7.5.

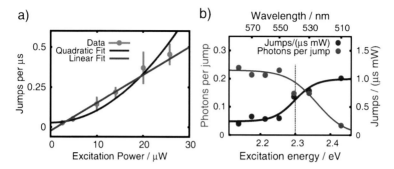

Figure 7.11 (a) Typical dependency of the spectral diffusion rate on the excitation power. The blue curve is a linear fit to the data with the jump rate at zero excitation power being a free fit parameter. The red curve is a quadratic fit to the data. Obviously, the linear function fits better to the data. (b) The number of collected photons from the ZPL transition per spectral jump as a function of the excitation photon energy. The blue data points were measured at equal photon count rates and varying excitation powers (left axis). Point size corresponds to the error bar. The blue curve is a guide to the eyes. The red points show the spectral jump rate normalized to the excitation power, calculated from the same data set (right axis). It is clearly visible that the spectral diffusion rate increases dramatically above the threshold of about 2.3 eV. From Ref. [154], © 2013 APS.

Excitation Power Dependency of Spectral Diffusion To gain further insight into the origin of the spectral diffusion, the dependency of the spectral diffusion rate γ_D on the excitation power is measured. Therefore, the above described measurement is repeated for several laser powers between 3 µW and 23 µW, well below the saturation power. The resulting data [Fig. 7.11(a)] show a clear linear dependency. Thereby two photon processes like the photo induced charge conversion [168] can be ruled out to be the origin of spectral diffusion. Remarkably for zero excitation power, the spectral diffusion rate reaches zero within the precision of the fit, although the intersection with the axis is a free fit parameter. From this, one can make two conclusions: *First, the excitation with the green laser is the main cause of spectral diffusion*. This is consistent with the simple charge trap model. The laser ionizes impurities, providing floating charges which can be trapped in other charge traps. *Second, the number of photons in the ZPL collected per spectral jump is constant* *for excitation far below saturation, i.e., independent of the excitation power.* This number directly evaluates the quality of the single-photon emission in terms of the possible number of subsequent indistinguishable photons available. This is a key figure of merit for future quantum optics experiments with NV centers. Obviously, in the linear regime of excitation power (far below saturation) the quality of single photon emission cannot be improved by reducing the excitation laser power, as it is often done in experiments with self-assembled quantum dots [169].

In order to derive a strategy to increase the number of collected photons per spectral jump its dependency on the temperature and excitation energy is investigated. Within the temperature range of 5–20 K no change in the spectral diffusion rate was observed. This is again consistent with the laser being the main cause of spectral diffusion by ionization of charge traps since their ionization energy exceeds $k_b T$ at this temperature range.

Excitation Wavelength Dependency of Spectral Diffusion In order to measure the influence of the excitation energy, the 532 nm cw laser is replaced with a pulsed *supercontinuum source* with a set of exchangeable 10 nm broad bandpass filters. While keeping the count rate in the ZPL constant at (1.4 ± 0.2) kcts/s (well below saturation) the spectral diffusion rate is measured for several excitation wavelength from 510 nm to 580 nm, corresponding to photon energies between 2.1 eV and 2.4 eV. The number of collected photons in the ZPL per spectral jump and the spectral jump rate normalized to the excitation power is plotted in Fig. 7.11(b).

The measurement clearly indicates that the number of collected photons per spectral jump decreases and spectral diffusion rate increases, respectively, with increasing excitation energy. Remarkably, there is a pronounced threshold of about 2.3 eV. This gives evidence for the existence of deep charge traps with an ionization energy close to 2.3 eV. The remaining spectral diffusion for energies below threshold might be attributed to substitutional nitrogen atoms in the diamond nanocrystals forming donor levels which are ionized at 1.7 eV [158, 159]. To identify the trap states as well as to generally modify these states in order to reduce spectral diffusion, more extensive studies including also higher quality diamond with natural or implanted defects must be performed. Potentially using two-color excitation, the studies can be extended to energies below the ZPL transition to investigate the influence of the nitrogen donors at 1.7 eV.

A promising strategy to maximize the number of subsequent indistinguishable single photons from a nanodiamond is to use relatively high excitation powers, but clearly below the saturation level and to chose a proper low excitation wavelength. Actively enhancing the optical transition strength [129, 131] or collection rate is also required, possibly to enable active line stabilization schemes, as reported for NV centers in higher quality bulk diamond [160]. Surface treatment accompanied by spectral diffusion studies should be extended to learn more about the dynamics of trap states on the diamond surface. When the outlined strategies succeed in controlling and minimizing spectral diffusion, NV centers in nanodiamond can be used as single-photon sources for linear optics quantum computing [170] or for the entanglement schemes discussed in Chapter 15.

7.5 Spin Physics of Nitrogen-Vacancy Centers

As noted at the beginning of this chapter, NV centers provide a triplet ground state. Today, this ground state is regarded as one of the most promising resources for quantum information technology. To understand the spin manipulation experiments presented in later sections, first the NV orbitals and triplet levels are reviewed in the following Section 7.5.1, while the influence of the singlet levels is discussed in Section 7.5.2. Remarkably, it is found that the latter ones give rise to a simple spin initialization and detection mechanism.

7.5.1 *Orbitals and Triplet Levels*

To obtain insight into the spin physics of the NV it is necessary to have a closer look on the orbitals and level structure [171–175]. In the following paragraphs this is done, mainly based on results from Ref. [175].

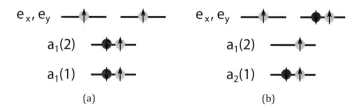

(a) (b)

Figure 7.12 Occupation of the electronic levels given by Eqs. 7.7–7.10 in a single particle picture. (a) In the NV ground state the a_1 states are completely filled, while e_x and e_y are occupied by electrons of parallel spin, resulting in a triplet character. (b) The $a_1(1)^2 a_1(2) e^3$ configuration of the first exited triplet-like state. The antisymmetric multi particle wave functions corresponding to these configurations are listed in Table 7.3.

NV Orbitals   The NV orbitals can be constructed as linear combinations of four sp^3 orbitals, labeled $|\sigma_1\rangle$, $|\sigma_2\rangle$, $|\sigma_3\rangle$, $|\sigma_N\rangle$. The first three correspond to the dangling bonds of the three carbon atoms neighboring the vacancy, while σ_N is associated with the nitrogen atom (Fig. 7.3(a)). From these orbitals four linear independent c_{3v}-symmetric wave functions can be constructed. These are

$$|a_1(1)\rangle = \alpha\,|a_c\rangle + \beta\,|\sigma_N\rangle, \tag{7.7}$$

$$|a_1(2)\rangle = \beta\,|a_c\rangle + \alpha\,|\sigma_N\rangle, \tag{7.8}$$

$$|e_x\rangle = (2\,|\sigma_1\rangle - |\sigma_2\rangle - |\sigma_3\rangle)/\sqrt{6}, \tag{7.9}$$

$$|e_y\rangle = (|\sigma_2\rangle - |\sigma_3\rangle)/\sqrt{2},\,, \tag{7.10}$$

where $|a_c\rangle = (|\sigma_1\rangle + |\sigma_2\rangle + |\sigma_3\rangle)/\sqrt{3}$ and the constants α and β are defined by the Coulomb interaction and satisfy $\alpha^2 + \beta^2 = 1$. As indicated by the names, $a_1(1, 2)$ transforms as the A_1 representation of the C_{3v} symmetry group, whereas $e_{x,y}$ transform as E, respectively [149, 173]. The energetic order of these orbitals can be found from quantitative electron-ion interaction models or density functional calculations to be $a_1(1)$, $a_1(2)$ and $e_{x,y}$, where the latter two are degenerate.

Triplet Wavefunctions   In the negatively charged NV center, these orbitals are occupied by six electrons. These can be arranged in total antisymmetric wave functions with symmetric and antisymmetric spin state, having triplet and singlet character respectively. Neglecting spin interactions, it turns out that the ground state is formed

Table 7.3 Triplet-like hole wave functions of the NV center

Configuration	State	Symmetry
	$^3A_{2-} = |e_x e_y - e_y e_x\rangle \otimes |\downarrow\downarrow\rangle$	$E_1 + E_2$
e^2	$^3A_{20} = |e_x e_y - e_y e_x\rangle \otimes |\downarrow\uparrow + \uparrow\downarrow\rangle$	A_1
	$^3A_{2+} = |e_x e_y - e_y e_x\rangle \otimes |\uparrow\uparrow\rangle$	$E_1 - E_2$
	$A_1 = |E_-\rangle \otimes |\uparrow\uparrow\rangle - |E_+\rangle \otimes |\downarrow\downarrow\rangle$	A_1
	$A_2 = |E_-\rangle \otimes |\uparrow\uparrow\rangle + |E_+\rangle \otimes |\downarrow\downarrow\rangle$	A_2
$a_1(2)e$	$E_1 = |E_-\rangle \otimes |\downarrow\downarrow\rangle - |E_+\rangle \otimes |\uparrow\uparrow\rangle$	E_1
	$E_2 = |E_-\rangle \otimes |\downarrow\downarrow\rangle + |E_+\rangle \otimes |\uparrow\uparrow\rangle$	E_2
	$E_y = |Y\rangle \otimes |\uparrow\downarrow + \downarrow\uparrow\rangle$	E_1
	$E_x = |X\rangle \otimes |\uparrow\downarrow + \downarrow\uparrow\rangle$	E_2

Hole wave functions of the first two manifolds with symmetric spin configuration. $\downarrow,\uparrow$ denotes the spin orientation of the hole, while $E_\pm = (e_x \pm i e_y) a_1(2) \pm a_1(2)(e_x \pm i e_y)$, $X = (E_- - E_+)/2$ and $Y = i(E_- + E_+)/2$. According to Ref. [175].

by the threefold degenerate triplet-like configuration $a_1(1)^2 a_1(2)^2 e^2$ as illustrated in Fig. 7.12(a), while the first excited triplet-like manifold has the six-fold degenerate configuration $a_1(1)^2 a_1(2)e^3$ (Fig. 7.12(b)). This triplet ground state is probably the most remarkable feature of the NV center, as unlike in most molecules it allows for spin experiments in the ground state.

The state notation simplifies in a *hole-picture*, where only the unoccupied states are noted. In this picture the ground state is simply e^2, while the first exited triplet-like state gets $a_1(2)e$. Table 7.3 summarizes these hole wave functions and their symmetry properties. The degeneracy of these multiplets is partially lifted by spin-orbit and spin-spin coupling. As illustrated in Fig. 7.13. At zero magnetic field, spin-orbit coupling splits the $m_s = \pm1$ ground states $^3A_{2\pm}$ from the $m_s = 0$ counterpart $^3A_{20}$ by approximately 2.9 GHz. Similar in the excited state the spacings from $E_{1,2}$ to $E_{x,y}$ to A_1 to A_2 are approximately 3.4 GHz, 4.4 GHz and 2.0 GHz respectively.

Optical Transitions The ground and excited states having equal spin components are linked by optical transitions near 637 nm, corresponding to the ZPL as illustrated in Fig. 7.13. Remarkably, here the $m_s = 0$ ground state $^3A_{20}$ forms a spin preserving cycling

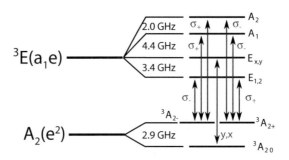

Figure 7.13 The electronic triplet levels of the NV center. The NV center exhibits a triplet ground state transforming as $^3A_2$ and an excited state ^3E between which optical transitions with indicated polarization occur at about 637 nm. The $m_s = 0$ state forms cycling transition with the $E_{x,y}$ manifold, while the other excited states couple to both $m_s = \pm 1$ ground states $^3A_{2\pm}$.

transition with the $E_{x,y}$ manifold. In contrast, the $E_{1,2}$ and $A_{1,2}$ excited state couple to both $m_s = \pm$ ground states $^3A_{2\pm}$, forming a λ-type scheme used in many experiments [121].

7.5.2 Singlet Levels and Spin State Detection

For spin state detection one optical transition, e.g., $^3A_{20}$ to E_x can be selectively driven, while detecting fluorescence from the phonon side band detected [150]. Alternatively, the NV can be excited off-resonantly, while the spin dependent energy of the subsequently emitted photon is detected. In both cases in principle a single excitation, decay, and photon detection cycle extracts enough information to unambiguously determine the spin state, i.e., perform a projective measurement. In a real experiment, there is no such on-demand projective measurement. For example a low photon collection and detection efficiency [39] requires repeated measurements. In case of the NV center a low Debye–Waller factor and spectral diffusion further impede the prediction of the system's spin state. Anyhow, inter system crossing to the NV singlet levels allows for another efficient spin detection mechanism as described in the following.

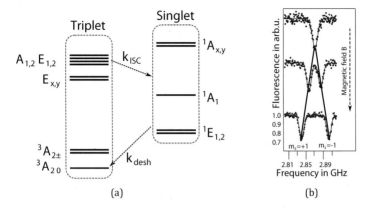

Figure 7.14 ISC processes and magnetic field splitting. (a) The NV center exhibits a triplet ground state transforming as $^3A_2$ and an excited states ^3E between which optical transitions occur. The metastable singlet states $^1A_1$ and ^1E can be reached nonradiatively via inter-system crossing (ISC). (b) In external magnetic fields, the degenerate $^3A_{2\pm}$ levels split up. This can be observed in ODMR measurements.

Inter System Crossing In addition to the triple state several singlet states with symmetries $^1A_1$ and $^1E_{1,2}$ exist, as illustrated in Fig. 7.14(a). These singlet states cannot be reached by optical transitions from the triplet, but might be occupied due to inter-system crossing (ISC) from the excited states. It appears, that these processes are unlikely for $E_{x,y}$, while comparable fast for the $A_{1,2}$ and $E_{1,2}$ manifold. Hence, starting with an $m_s = \pm 1$ ground state, ISC will appear likely after few optical cycles. As the deshelving occurs exclusively to the $m_s = 0$ ground state, this allows for very efficient spin polarization, even by off-resonant excitation. Furthermore, as the dwell time in the optically inactive singlet manifold is on the order of 10 times the excited state lifetime, the $m_s = \pm 1$ ground state appear darker in fluorescence measurements. These unique features of spin initialization and readout render the NV center ideal for spin experiments.

7.5.3 Optical Detection of Magnetic Resonances

Probably the most fundamental spin manipulation experiment on NV centers is the optical detection of magnetic resonances (ODMR):

The NV center is weakly excited off-resonantly, while a magnetic AC field is applied. In experiments, this is frequently done by near-field coupling to a thin (~30 μm) gold wire, or microwave (MW) guides fabricated directly on the sample. When the magnetic field gets in resonance with the level splitting between $m_s = 0$ and $m_s = \pm 1$, spin flips occur. This goes in hand with a drop in fluorescence, being easily detected. Using this scheme, it is comparably easy to detect the ground state's magnetic resonance at 2.9 GHz.

While at zero magnetic field the $m_s = 0 \leftrightarrow m_s = \pm 1$ transitions are degenerate, this changes in magnetic fields where the Zeeman effect lifts the degeneracy. This makes the electronic spin of the NV center ideal to measure magnetic fields on the single spin level [135, 176–181]. The sensitivity of such magnetometers is mainly limited by the line width of the transition. Here, an ultimate limit is given by the coherence time T_2 of the transition, which is determined by magnetic field fluctuations induced by spins in the vicinity as discussed in Section 4.1.6. In diamond which is mainly spin-free such fluctuations are only very weak and T_2 reaches the order of several microseconds. To observe this limit, power broadening by either optical excitation or the microwave drive itself has to be excluded.

This can be done using pulsed laser schemes and T_2^* limited lines are easily observed, where the resolution can be further enhanced toward the ultimate T_2-limit using spin-echo techniques. By high resolution spectroscopy, it can be observed, that the $m_s = 0 \leftrightarrow m_s = \pm 1$ are themselves split by about 2 MHz into hyperfine triplets, originating from coupling to the ^{15}N nuclear spin of the NV center. This coupling provides access also to nuclear spins which is exploited in many experiments via ENDOR, e.g., Refs. [32, 90, 122, 182–188].

7.5.4 Coherent Spin Manipulation

Probably the most fundamental coherent spin manipulation experiment is to drive Rabi oscillations. Owing to the possibility of optical spin initialization and readout, and the long spin coherence time, it can be easily done on NV centers. For this, the NV spin must be initialized to $m_s = 0$ using excitation with 532 nm for several μs using a setup as illustrated in Fig. 7.15. About 1 μs after switching

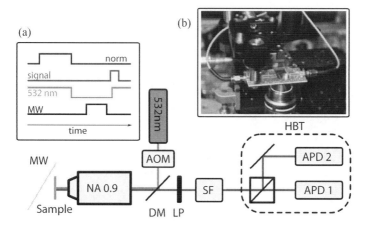

Figure 7.15 Optical setup for ODMR measurements on NV centers. Similar to the previously shown setup of Fig. 7.4 single NV centers are excited through an objective lens with high numerical aperture (NA 0.9). Here, the excitation laser can be switched by an *acousto-optic modulator* (AOM). Fluorescence light is collected through the same objective lens and separated by a dichroic mirror (DM). Residual laser light is filtered by a long-pass filter (LP), while stray light is removed by a spatial filter (SF) prior detection by a Hanbury Brown and Twiss interferometer (HBT). Microwaves can be applied by near field coupling to a thin ($\sim$30 µm) gold wire (MW). A fast pattern generator is used to generate measurement sequences. Inset (a) shows a sequence to measure Rabi oscillations. The NV center is initialized by a 532 nm laser pulse, a microwave pulse of variable length is applied and the laser switched on again for readout. Measurement signal and normalization reference acquisition are gated as indicated. Inset (b) is a photograph of the setup showing the sample, the microwave antenna and the objective lens. From Ref. [189] © 2013 Nature Publishing Group.

of the laser, the NV is relaxed from the singlet to the desired $m_s = 0$ triplet ground state. Here, a microwave pulse of defined length can be applied to drive Rabi oscillations. To read out the final spin state, the laser is switched on again and the fluorescence is detected. Because of the ISC mechanism described earlier in the first $\sim$500 ns after switch-on, the $m_s = \pm 1$ states appear darker than their $m_s = 0$ counterpart and thereby can be distinguished. To account for intensity fluctuations due to drift of the sample during the data acquisition, it is convenient to record a normalization reference signal at the end of the initialization process.

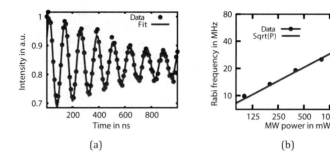

(a)   (b)

Figure 7.16   Rabi oscillations of the NV's electronic spin. (a) The fluorescence intensity of a NV center after application of a MW pulse in resonance with the $m_s = 0 \Leftrightarrow m_s = +1$ transition of varying length. The contrast is about 30%, and the coherence time T_2, after which the oscillation amplitude decreases to $1/e$ in 0.8 μs. (b) The Rabi frequency as a function of the MW power P. The measurement perfectly reproduces the theoretically predicted square root dependency.

Figure 7.16 shows a typical example of Rabi oscillations measured on a single NV center in nanodiamond. The Rabi oscillation decay time is on the order of about 1 μs, and the achieved contrast is 30%. This allows to measure many oscillations at Rabi frequencies on the order of 10–100 MHz. The Rabi frequency is proportional to the square root of the microwave power (cf. Fig. 7.16(b)), i.e., depends linearly on the magnetic field strength as expected from theory. To increase the Rabi frequency, centers closer to the microwave antenna or with more favorable orientation can be used. Alternatively, the NV centers can be placed in microwave resonators, or microscopic antennas might be lithographically defined in the close proximity of the NV center.

7.6 Simplified Model and Effect of Strain on Nitrogen-Vacancy Centers

Most experiments use only the $m_s = 0$ and $m_s = -1$ ground states. In this case, the third ground state $m_s = +1$ can be neglected and only the former ones need to be considered. Furthermore, at room temperature, transition to the excited states cannot be resolved spectroscopically. Thus, to obtain a simplified model, the excited

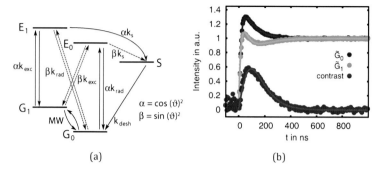

(a)   (b)

Figure 7.17 Spin dynamics of NVs with strain. (a) Simplified level diagram of the NV with strain. The $m_s = 0$ (1) electronic ground state is denoted $G_{0(1)}$, while the corresponding predominantly $m_s = 0$ (1) excited states are $E_{0(1)}$. The singlet states are merged to the state S. Solid and dashed arrows correspond to allowed and forbidden transitions, respectively. Coherent ground state spin rotations are indicated by MW. (b) Fluorescence intensities $I_{G_{0(1)}}(t)$ of a NV center after switch-on of the cw laser when either the predominantly $m_s = 0$ state $\tilde{G}_0$ or the $m_s = 1$ state $\tilde{G}_1$ was initially prepared (see text). The contrast defined by $(I_{G_0} - I_{G_1})/(I_{G_0} + I_{G_1})$ is scaled up by a factor of 2 for better visibility. The solid curves are numerical solutions of the rate equation corresponding to the level diagram depicted in (a). Adapted from Ref. [190].

states linked to each of the considered two ground states, as well as the single states, can be merged into one single state each [191]. This results in a five-level model where the $m_s = 0$ and $m_s = 1$ ground states are labeled G_0 and G_1, while their excited state counterparts are labeled $E_{0(1)}$, respectively. For real NV centers this scheme is however incorrect, as strain induces a mixing of the exited states. To account for this, the spin mixing angle ϑ can be introduced phenomenologically. This situation is illustrated in Fig. 7.17(a). From this model a reduced polarization efficiency of the ground state spin η_{pol} is expected. By illumination with green light the bright state $\tilde{G}_0 = \eta_{pol}G_0 + (1 - \eta_{pol})\exp(i\varphi_1)G_1$ with random phase φ_1 is prepared.

The mixing has two direct implications: spin nonpreserving optical transitions, as well as inter-system crossing from the predominantly $m_s = 0$ excited state become possible [124, 171]. According to Fermi's golden rule, the rates of the transition allowed in the unstrained NV are reduced by the factor $\alpha = \cos(\vartheta)^2$, whereas the formally forbidden transitions have now the rates $k_x\beta = k_x \sin(\vartheta)^2$, with k_x being the rate of the corresponding allowed transitions.

As $m_s = -1$ is not considered, the system can be described by the vector $x = \begin{pmatrix} G_0 & G_1 & \Im(C_{01}) & E_0 & E_1 & S \end{pmatrix}^T$, where the entries correspond to the level populations, while $\Im(C_{01})$ denotes the imaginary part of the coherence between G_0 and G_1. In this notation, the dynamics of x is given by $\frac{d}{dt}x = Ax$, with

$$A = \begin{pmatrix} -k_{exc} & 0 & -\Omega & \alpha k_{rad} & \beta k_{rad} & k_{desh} \\ 0 & -k_{exc} & \Omega & \beta k_{rad} & \alpha k_{rad} & 0 \\ \frac{i\Omega}{2} & -\frac{i\Omega}{2} & \Gamma & 0 & 0 & 0 \\ \alpha k_{exc} & \beta k_{exc} & 0 & -k_{rad} & 0 & \beta k_S \\ \beta k_{exc} & \alpha k_{exc} & 0 & -k_{rad} & 0 & \alpha k_S \\ 0 & 0 & 0 & \beta k_S & \alpha k_S & -k_{desh} \end{pmatrix} \tag{7.11}$$

and $\Gamma = -(1/T_2^* + k_{exc})$.

In most NV centers transition rates and parameters vary due to different strain conditions and orientations. To estimate the parameters for a particular single NV, characterization can be performed with the setup shown in Fig. 7.15. First the ground state spin resonances, which are split up by a small permanent magnetic field by approximately 200 MHz, are identified [118]. Subsequently, coherent Rabi oscillations are driven to measure the oscillation frequency, as well as the damping of the oscillation $1/T_2^*$. Here, the Rabi frequency must be chosen to be much smaller than the splitting between the $m_s = \pm 1$ levels in order to allow for individual addressing of the $m_s = 0$ to $m_s = +1$ transitions. To finally estimate the transition rates between different levels, first the $m_s = 0$ state is prepared by applying a green cw laser (~ 0.73 mW) for about 5 µs. About 1 µs after switching off of the laser the NV is assumed to be relaxed into the desired $m_s = 0$ state. Optionally, it can be transferred into the $m_s = 1$ state by an additional MW π-pulse. Subsequently, the time-dependent fluorescence intensity after switching on the cw laser again is measured (Fig. 7.17(b)).

These dynamics strongly depend on the transition rates, allowing to deduce all free parameters from a fit. Table 7.4 shows typical parameters of a NV center in type Ia bulk diamond. For additional

Table 7.4 Transition rates and parameters of a typical NV.

$2\pi/\Omega$	(240 ± 7) ns	T_2^*	(0.5 ± 0.1) μs
$1/k_{exc}$	(30.5 ± 5) ns	$1/k_{rad}$	(13 ± 4) ns
$1/k_{desh}$	(220 ± 60) ns	$1/k_S$	(15.4 ± 5) ns
ϑ	$(12.4 \pm 3)°$	I_{bg}	0.2 ± 0.1
η_{pol}	0.92 ± 0.01		

Transition rates and parameters deduced by fitting the model Fig. 7.17(a) to the measurement shown in Fig. 7.17(b). The error corresponds to one standard deviation confidence interval.

verification, the excited states lifetimes can be independently measured with a pulsed laser, the excitation rate k_{exc} can be obtained from saturation measurements, and the contribution of fluorescent background to the signal I_{bg} can be estimated from autocorrelation measurements.

From the model the value of the ground state polarization efficiency η_{pol}, i.e., the form of the state $\tilde{G}_0$ can be deduced. A subsequent microwave π-pulse transfers the population into the dark state $\tilde{G}_1 = \eta_{pol}G_1 + (1 - \eta_{pol}) \exp(i\varphi_2)G_0$.

While the described simple model is sufficient to describe ground state spin experiments, a more accurate model must be used to describe experiments involving excited state spin manipulation, or low-temperature experiments, where the individual transitions can be resolved [121, 124, 172, 193]. For this, the eigenvalues of the

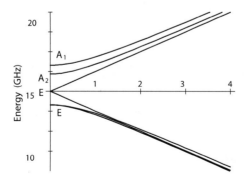

Figure 7.18 *Breit–Rabi diagram* showing the excited state eigenvalues as a function of strain in GHz. Already small amounts of strain lift the degeneracy between E_x and E_y. Adapted from Ref. [192]

excited state Hamiltonian can be calculated [174, 175, 192, 194]. One finds that strain lifts the degeneracy of the $E_{1,2}$ and $E_{x,y}$ levels as illustrated in Fig. 7.18. Furthermore, a level mixing occurs, allowing spin nonpreserving optical transitions between the $m_s = 0$ and $m_s = \pm 1$ manifold. This can be used to identify the NV center with a three-level λ-type scheme as demonstrated in several experiments on all-optical spin manipulation [124, 126, 127, 195].

7.7 Demonstration of the Quantum Zeno Effect

As described above, optical spin detection and initialization combined with long coherence times in the mainly spin-free diamond lattice render the NV ideal to demonstrate coherent spin manipulation. For example, electromagnetic-induced transparency [118], simple quantum algorithms [119], and subdiffraction optical magnetometry [196] have been demonstrated. Here, an experiment on the quantum Zeno effect (Section 4.2.9) will be discussed on the basis of Ref. [190].

As discussed in Section 4.2.9, the quantum Zeno effect can be witnessed by the inhibition of coherent population transfers by measurements. Today, this is interpreted as measurement-induced decoherence. To demonstrate this on NV centers, it is important to note that in the NV center the measurement process is unavoidably accompanied by repolarization of the NV center. Thus, it is not sufficient to show that a population transfer from the $m_s = 0$ to $m_s = +1$ can be inhibited by measurement [118, 196], but also transfers in the reverse direction are affected. In the reverse direction, i.e., for the transition from $m_s = 1$ to $m_s = 0$, an inhibition cannot be explained by repumping and thus unambiguously demonstrated the quantum Zeno effect.

Experimental Setup For the experiment, the setup shown in Fig. 7.15 and the NV in type Ib bulk diamond with a solid immersion lens (SIL) produced by focused ion beam milling [150] from the measurement shown in Fig. 7.17 are used. As here the NV parameters are known, the quantum Zeno experiment cannot only be experimentally realized but also simulated. The experimental

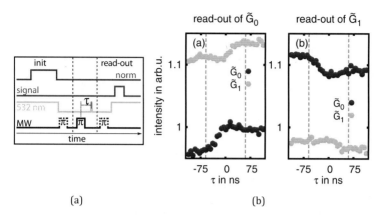

read-out of $\tilde{G}_0$ read-out of $\tilde{G}_1$

(a)   (b)

Figure 7.19 (a) Pulse sequence used to measure the Zeno effect. MW denotes the microwave source applying up to three π-pulses. The 532 nm cw laser initializes the NV at the beginning of the experiment and performs state readout at the end, while a 12 ns short pulse is applied at time delay τ with respect to the center of the central π-pulse. The NV fluorescence intensity is measured within short time windows after and prior to applying the π-pulses to obtain the signal and normalization reference. (b) Measurement of the NV fluorescence after the quantum Zeno experiment. The NV is prepared to be in the bright $\tilde{G}_0$ (dark $\tilde{G}_1$) state. The coherent population transfer during the MW pulse (indicated by dashed lines) is inhibited by a short green laser pulse at time delay τ with respect to the center of the MW pulse. Subsequently, the fluorescence as a measure of the occupation of the bright state $\tilde{G}_0$ is probed. The Zeno pulse is most effective at $\tau = 0$. The right panel shows the same, but with an additional π-pulse before measuring the NV fluorescence, i.e., probing of the dark state $\tilde{G}_1$. Adapted from Ref. [190].

sequence is illustrated in Fig. 7.19(a). First, the NV is initialized to the bright state $\tilde{G}_0$ by applying the green laser for about 5 μs. To prepare the dark $\tilde{G}_1$ state a subsequent MW π-pulse can be applied. After the initialization, the MW pulse is switched on starting a coherent transition from $\tilde{G}_0$ to $\tilde{G}_1$ (or from $\tilde{G}_1$ to $\tilde{G}_0$). The pulse is set to a fixed length of 120 ns, i.e., a π-pulse. Synchronized to the microwave pulse, a short laser pulse (pulse length 18 ns, peak power 730 μW) is applied at varying time delay τ. With about 30% probability this initiates a single cycle of excitation, subsequent spontaneous emission, and possible state-selective photon detection, i.e., a measurement of the NV center's spin state. Even if the final state-selective detection is only done in principle, the ground state coherence is destroyed effectively. Finally, about 300 ns after the microwave pulse the green cw laser is turned on again and the fluorescence of the NV is recorded, giving a measure of the remaining population of the brighter $\tilde{G}_0$ state.

Independent of the microscopic mechanism, which might be phonon coupling, ISC, or the photon emission and its subsequent absorption, the laser initiates a measurement of the spin state, destroys the microwave-induced coherent polarization, and thereby inhibits the dynamics, similar to experiments with ions [39].

Experimental Results In the experiment shown, the inhibition efficiency depends on the time delay between the laser and the MW pulse. At the center of the MW pulse the polarization reaches its maximum. Hence a projective laser pulse at $\tau = 0$ effectively inhibits further coherent dynamics and the *final* state has a large component of the *initial* state. This behavior is clearly visible in Fig. 7.19(b) as increased (decreased) fluorescence around $\tau = 0$ when initially $\tilde{G}_0$ ($\tilde{G}_1$) was prepared. In particular, for an initial $\tilde{G}_1$ the decreasing intensity at $\tau = 0$ in Fig. 7.19(b) proves that the effect is not due to repumping of the NV center. In this configuration the laser pulse effectively *increases* the population of the $\tilde{G}_1$ state, resulting in a decreased fluorescence intensity when probing. Furthermore, owing to the large ratio between MW pulse length and excited state lifetime, excitation into the electronic excited state where microwave pulses are off resonant can be excluded as cause of the effect.

To get further insight, the experiment is repeated with an additional π-pulse before measuring the NV fluorescence (Fig. 7.19(b) right panel) and the spin projection $< m_s >$ is calculated from the measured contrast between the intensity with and without this final π-pulse. Figure 7.20 shows this spin projection together with numerical simulations based on the simple strained NV model of Fig. 7.17(a). After preparation of the initial state the polarization efficiency is $\eta_{pol} = 0.92$. Thus, the contrast after preparation of $\tilde{G}_0$ corresponds to $< m_s >= 0.08$, while for the achieved contrast after preparation of $\tilde{G}_1$ it is assumed that $< m_s >= 0.92$. In an ideal Zeno experiment, where the measurement completely destroys

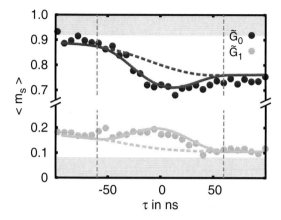

Figure 7.20 Experimental values (dots) and numerical simulation (solid curves) of the m_s spin projection when initially preparing the predominantly $m_s = 0$ (upper dataset) and $m_s = 1$ (lower dataset) level, respectively. Time interval between the vertical lines corresponds to the MW pulse (see Fig. 7.19(a) for the measurement sequence), while the shaded areas are inaccessible because of imperfect polarization after initialization. The features centered at $\tau = 0$ are whiteness of the Zeno effect. Dashed lines indicate the theoretical expectations when decoherence due to the Zeno effect is artificially turned off. Adapted from Ref. [190].

the coherence without altering the populations, a detection pulse not overlapping with the MW pulse has no influence. A detection within the MW pulse, however, inhibits the coherent dynamics. At the extreme case $\tau = 0$ the coherent spin transfer is stopped and the final state in an ideal Zeno experiment is a 50/50 mixture of the two spin states.

Interpretation In the presented experiment the excitation probability is 30%, well below unity, and hence the efficiency of coherent spin transfer is reduced by about 10% only. Further deviations from the ideal case have several reasons. There is depolarization via spin nonpreserving transitions and repolarization via ISC, as well as the finite lifetime of the excited states. Furthermore, the limited dephasing time T_2^* damps the Rabi oscillations, resulting in a small offset toward $< m_s > = 0.5$, particularly visible at the curve for an initial $\tilde{G}_1$ for positive τ. While in the measurement depolarization via spin nonpreserving transitions can be neglected, repolarization via ISC is effective when the NV state has a large G_1 component at time τ. This drives the spin toward $< m_s >= 0$ for positive (negative) τ in case of initial preparation of $\tilde{G}_0$ ($\tilde{G}_1$). The finite lifetime of the NV excitation has to be considered when the laser pulse is applied before the end of the microwave pulse, i.e., for $\tau < 60$ ns. The laser pulse drives the NV to the excited states, where the MW pulse is ineffective. However, as the MW pulse is long compared to the excited state lifetime and the overall excitation probability is only about 30%, this effect can be neglected.

To support this analysis a simulation of the experiment is performed, where an excited state coherence term is artificially introduced to turn off the Zeno effect, while all other effects remain unaffected. While the simulation including the Zeno effect (solid line in Fig. 7.20) reproduces the measured data very well, the simulation without Zeno effect (dashed lines in Fig. 7.20) does not explain the feature centered at $\tau = 0$. This feature is caused by the destruction of the quantum coherence induced by the green laser pulse initiating the measurement of the spin state, i.e. the quantum Zeno effect.

Under spin-selective optical excitation, e.g., at cryogenic temperature this effect will not only be more pronounced but is also relevant for recently proposed robust two-qubit quantum gates [44–46]. Furthermore, as single defect centers are utilized as a scanning quantum emitter probe [129, 197, 198], where modifications of rates (lifetime, decoherence, etc.) by the local environment are monitored as probe signals. Here, detailed knowledge of the intricate interplay between coherent and incoherent dynamics paves the way to exploit full information on the spin dynamics. Thus much deeper insight into interactions of a single quantum system with its mesoscopic environment might be obtained.

PART III

OPTICAL MICROSTRUCTURES

Introduction

Today, microfabrication techniques allow to manufacture structures that control the propagation of light on the nanoscale. Such nanophotonic devices can be used to tailor the properties of single-quantum emitters, as discussed at the end of Part I. Thus, they can be used for many interesting quantum optics experiments and might become a basis for a future integrated quantum technology. Furthermore, integrated semiconductor optics will enable tremendous progress in VLSI electronic circuits. Currently, progress is limited by the broadband wiring of the different areas on a chip. For this problem, optical interconnections are a favorable solution.

Part III gives an introduction to the field and explains how dielectric structures can be used to tailor the flow of light on the nano- and microscale. Starting with Maxwell's equations derived in Part I, the classical theory of electrodynamics in dielectrics is developed. Applying this theory to the case of a uniform medium, modified expressions for the electric field per photon and the spontaneous emission rate are obtained. Later on, Maxwell's equations are translated into an eigenvalue problem, leading to the concept of optical Bloch waves and total internal reflection. The latter gives rise to a brief discussion of immersion microscopy as well as light guiding and confining structures, namely strip waveguides, fibers, and disk resonators. These resonators are a first example of the high-Q, low-mode volume cavity structures postulated in the discussion of cavity QED at the end of Part I. Here, fundamental experiments with such resonators are presented.

At the end of this part, more complex dielectric structures, the photonic crystals, are introduced. Here light is confined and guided by photonic bandgaps arising from periodic dielectric arrangements, in close analogy to solid state physics. After introducing the theoretical concepts and experimental methods, this part closes with a chapter on applications of photonic crystals. There, narrow-band filters, refractive index sensing, and thermo-optical switching are discussed with of numerical simulations and novel experimental results.

Chapter 8

Electrodynamics in Media

Part I treated the interaction of the modes of the electromagnetic field with single or a few individual quantum systems in empty space. While this is clearly appropriate for single atoms or ions trapped in vacuum chambers, it is definitely not valid for the solid state quantum emitters introduced in Part II. All of these emitters are embedded into some solid matrix, consisting of many individual atoms bound together. Even when the matrix interacts only weakly with the electromagnetic modes of interest, it will introduce photon scattering, i.e., mixing of different k vectors.

Fortunately it is sufficient to treat all atoms of the matrix and their interaction with the electromagnetic field classically to describe this mode mixing. Thus, at first a classical description of the electromagnetic field in media based on Maxwell's equations is developed in the following.

8.1 Maxwell's Equations in Dielectric Media

In the theory of electrodynamics in dielectric media, the material background is assumed to be a charge distribution which is neutral in average. In this background, externally applied fields $\mathbf{E}_{ext}$, $\mathbf{B}_{ext}$

Integrated Quantum Hybrid Systems
Janik Wolters
Copyright © 2015 Pan Stanford Publishing Pte. Ltd.
ISBN 978-981-4463-82-9 (Hardcover), 978-981-4463-83-6 (eBook)
www.panstanford.com

generated by the charge distribution ρ_{ext} and the current density $\mathbf{j}_{ext}$ might induce carrier dynamics, which again contribute to the total charge ρ_{tot}, current $\mathbf{j}_{tot}$, and electromagnetic field ($\mathbf{E}_{tot}$, $\mathbf{B}_{tot}$) according to Maxwell's equations. So, for example, a charge $\rho_{ind}(t)$ and hence a field $\mathbf{E}_{ind}$ can be a consequence of a small separation between electrons and nuclei as response to the overall electric field. When these charge distributions change, this is linked via the continuity equation to a current $\nabla \cdot \mathbf{j}_{ind} = -\dot{\rho}_{ind}(t)$ that induces the field $\mathbf{B}_{ind}$. In addition to $\mathbf{j}_{ind}$, the magnetic field might induce microscopic eddy currents $\mathbf{j}_{curl}$ that are not related to a change in the charge density, but generate the field $\mathbf{B}_{curl}$.

To describe this situation, one makes use of the linearity of Maxwell's equations and defines two quantities proportional to the induced fields. The *polarization* $\mathbf{P}$ is given by

$$\mathbf{P} = -\varepsilon_0(\mathbf{E}_{ind}), \tag{8.1}$$

and the *magnetization* $\mathbf{M}$ is given by

$$\mathbf{M} = \frac{1}{\mu_0}(\mathbf{B}_{ind} + \mathbf{B}_{curl}). \tag{8.2}$$

Rewriting Eq. 2.36 in terms of the polarization $\mathbf{P}$ yields

$$\nabla \cdot \mathbf{P} = -\rho_{ind}, \tag{8.3}$$

while the curl of the magnetization is given by Eq. 2.39 and the continuity equation to be

$$\nabla \times \mathbf{M} = \mathbf{j}_{curl}. \tag{8.4}$$

Using the linearity of electrodynamics and the definitions of $\mathbf{P}$ and $\mathbf{M}$ in Eqs. 8.1 and 8.2, the so-called *electric displacement field* $\mathbf{D} = \epsilon_0 \mathbf{E}_{ext}$ generated by the charge ρ_{ext} can be expressed as the sum of polarization and total electric field $\mathbf{E}_{tot}$

$$\mathbf{D} = \varepsilon_0 \mathbf{E}_{tot} + \mathbf{P}. \tag{8.5}$$

Similar, the $\mathbf{H}$-*field* defined as $\mathbf{H} = 1/\mu_0 \mathbf{B}_{ext}$ (where $\mathbf{B}_{ext}$ is generated by $\mathbf{j}_{ext}$) can be expressed as

$$\mathbf{H} = \frac{1}{\mu_0}\mathbf{B}_{tot} - \mathbf{M}. \tag{8.6}$$

The behavior of these fields **D** and **H** can be computed from Maxwell's equations Eqs. 2.36 and 2.39 for the total fields, the continuity equation, and the above relations Eqs. 8.3 and 8.4:

$$\nabla \cdot \mathbf{D} = \rho_{free}, \tag{8.7}$$

$$\nabla \times \mathbf{H} = \dot{\mathbf{D}} + \mathbf{j}_{free}. \tag{8.8}$$

These two equations together with Eq. 2.37 and Eq. 2.38 are usually called macroscopic Maxwell's equations and completely describe the field dynamics when induced charges and currents are present.

8.2 Linear Isotropic Dielectrics

By now the distinction between external and induced contributions has not really simplified the overall dynamics, since the magnetization **M** and polarization **P** are both unknown. Nevertheless, if the fields are not too strong, i.e., in the limit of linear optics, it is justified to assume that ρ_{ind} and $\mathbf{j}_{curl}$ are a linear and isotropic material response to the total electric and magnetic field, respectively, with

$$\mathbf{P} = \varepsilon_0 (\varepsilon_r - 1)\mathbf{E}_{tot}, \tag{8.9}$$

$$\mathbf{M} = \left(\frac{1}{\mu_0} - \frac{1}{\mu_0 \mu_r} \right) \mathbf{B}_{tot}, \tag{8.10}$$

the above equations Eqs. 8.5 and 8.6 simplify to

$$\mathbf{D} = \varepsilon_0 \varepsilon_r \mathbf{E}_{tot}, \tag{8.11}$$

$$\mathbf{H} = \frac{1}{\mu_0 \mu_r} \mathbf{B}_{tot.}, \tag{8.12}$$

with the *relative permittivity* ε_r and the *relative permeability* μ_r. In the absence of external charges and currents Maxwell's equations get the form

$$\nabla \cdot (\varepsilon_r(\mathbf{r})\mathbf{E}(\mathbf{r}, t)) = 0, \tag{8.13}$$

$$\nabla \cdot \mathbf{H}(\mathbf{r}, t) = 0, \tag{8.14}$$

$$\nabla \times \mathbf{E}(\mathbf{r}, t) + \mu_0 \dot{\mathbf{H}}(\mathbf{r}, t) = 0, \tag{8.15}$$

$$\nabla \times \mathbf{H}(\mathbf{r}, t) - \varepsilon_0 \varepsilon_r(\mathbf{r})\dot{\mathbf{E}}(\mathbf{r}, t) = 0. \tag{8.16}$$

Here the permittivity is explicitly allowed to be space dependent, corresponding to different materials being at different positions in space.

8.2.1 Electric Field per Photon

In most dielectric materials like semiconductors, charges are immobile and currents cannot be induced; thus $\mu_r \approx 1$. For vacuum the relative permittivity is unity, $\varepsilon_{vac} = 1$. In this case, Eqs. 8.11 and 8.12 are equal to Eqs. 2.37 and 2.38. The relative permittivity of most other materials is larger than unity, $\varepsilon_r > 1$. This expresses the fact that the induced electric field $\mathbf{E}_{ind}$ counteracts, i.e., effectively reduces the externally applied field. Hence in dielectric materials also the electric field per photon Eq. 2.6 is reduced to

$$\mathcal{E}_l = \sqrt{\frac{\hbar \omega_l}{2\varepsilon_0 \varepsilon_r(\mathbf{r}) L^3}}. \tag{8.17}$$

8.2.2 The Classical Wave Equation

Dividing Eq. 8.16 by $\varepsilon_r(\mathbf{r})$, taking the curl, and inserting the time derivative of Eq. 8.15 results in an equation for the **H**-field that is decoupled from the **E**-field:

$$\nabla \times \left[\frac{1}{\varepsilon_r(\mathbf{r})} \nabla \times \mathbf{H}(\mathbf{r}, t) \right] - \varepsilon_0 \mu_0 \frac{\partial^2}{\partial t^2} \mathbf{H}(\mathbf{r}, t) = 0. \tag{8.18}$$

This partial differential equation can be regarded as the fundamental problem of electrodynamics. The following sections deal with its solutions.

8.3 Spontaneous Emission in Uniform Dielectrics

In the case of a homogenous material covering the entire space, i.e., $\varepsilon_r(\mathbf{r}) = const$ the relation $\nabla \times (\nabla \times \mathbf{a}) = \nabla(\nabla \cdot \mathbf{a}) - \nabla^2 \mathbf{a}$ and Eq. 8.14 can be used to transform Eq. 8.18 into a wave equation. Transforming to Fourier space, this reads

$$\mathbf{k}_l^2 \tilde{\mathbf{H}}_l(t) - \varepsilon_0 \varepsilon_r \mu_0 \frac{\partial^2}{\partial t^2} \tilde{\mathbf{H}}_l(t) = 0. \tag{8.19}$$

Obviously the Fourier components of the **H** field behave like harmonic oscillators obeying the dispersion relation

$$\omega_l = \frac{c}{n} |\mathbf{k}_l|, \tag{8.20}$$

with the vacuum speed of light $c = 1/\sqrt{\varepsilon_0\mu_0}$ and the *refractive index* $n = \sqrt{\varepsilon_r}$. The same dynamics and dispersion relations could have been derived for the vector potential **A**, the transverse electric field $\mathbf{E}_T$, and hence also for the normal modes α. Thus, in the case of uniform dielectrics, the canonical quantization (cf. Section 2.3) can be followed with the modified dispersion relation and constant $\mathcal{E}_l$.

The modifications given by Eqs. 8.17 and 8.20 compared to the free space situation have important consequences for the spontaneous emission rate [Eq. 4.23] in dielectric media. At first glance the electric field per photon is reduced, resulting in reduction of the spontaneous emission by $1/\varepsilon_r$. Nevertheless, this is compensated by an increase in the density of states by n^3 due to the modified dispersion relation (cf. Section 3.4). In total this results in an increased spontaneous emission rate in the dielectric by a factor of n:

$$\gamma_{spon}^{diel} = n\gamma_{spon}, \tag{8.21}$$

where the spontaneous emission rate in the vacuum γ_{spon} is given by Eq. 4.23. This in principle enhances also the photon collection rate obtained from a single emitter Eq. 4.22. Nevertheless, in the case that not the entire space is covered with the dielectric, the enhancement in spontaneous emission can eventually not be exploited as demonstrated later on.

8.4 Electrodynamics as an Eigenvalue Problem

While in the previous case of a uniform dielectric the uncoupled harmonic normal modes could easily be obtained by scaling the dispersion relation and the electric field per photon, this is a complicated task when arbitrary distributions of dielectrics are treated. One approach to tackle this problem is to reformulate Eq. 8.18 as an eigenvalue problem, where the normal modes can be found as eigenvectors, while the eigenvalues correspond to the frequencies of the modes.

For this, harmonically oscillating complex modes of the form

$$\mathbf{H}(\mathbf{r}, t) = \mathbf{H}'(\mathbf{r})e^{-i\omega t}, \tag{8.22}$$

are assumed. Here, only the real part of the complex modes has a physical meaning, corresponding to the **H** field. Inserting the Ansatz into the wave equation Eq. 8.16 gives the *master equation of electrodynamics* [199]:

$$\hat{\Theta}\mathbf{H}'(\mathbf{r}) = \left(\frac{\omega}{c}\right)^2 \mathbf{H}'(\mathbf{r}), \tag{8.23}$$

with the linear operator $\hat{\Theta}$ defined via

$$\hat{\Theta}\mathbf{H}'(\mathbf{r}) = \nabla \times \left[\frac{1}{\varepsilon_r(\mathbf{r})}\nabla \times \mathbf{H}'(\mathbf{r})\right]. \tag{8.24}$$

One can show [199] that $\hat{\Theta}$ is Hermitian, i.e., $(\hat{\Theta}\mathbf{H}_1, \mathbf{H}_2) = (\mathbf{H}_1, \hat{\Theta}\mathbf{H}_2)$, with the inner product defined by

$$(\mathbf{H}_1, \mathbf{H}_2) = \int d^3\mathbf{r}\mathbf{H}_1^* \cdot \mathbf{H}_2. \tag{8.25}$$

From the hermiticity of $\hat{\Theta}$ one can conclude that all eigenvalues ω are real and that different eigenmodes are orthogonal,[1] i.e., are decoupled. Hence with proper normalization the variables P and Q associated with the eigenmodes can serve as a quantization basis.

8.5 Symmetries in Dielectric Structures

It is well known that symmetries often significantly simplify physical problems and determine many properties of the symmetric system. Fortunately, this is also the case for the modes of the electromagnetic fields. Here, the important cases of *mirror symmetry*, *continuous translational symmetry*, and *discrete translational symmetry* are considered.

8.5.1 *Mirror Symmetries*

A dielectric configuration given by $\epsilon_r(\mathbf{r})$ is symmetric under inversion at a plane with normal vector $\mathbf{e}_p$ if

$$\epsilon_r(\mathbf{r}) = \hat{\mathcal{M}}\epsilon_r(\mathbf{r}), \tag{8.26}$$

---

[1] In case of degenerate modes, it is always possible to find the orthogonal basis of the degenerate subspace.

with the *mirror operator* $\hat{\mathcal{M}}_{\mathbf{e}_p}$ defined by

$$\hat{\mathcal{M}}_{\mathbf{e}_p}\mathbf{a} = \hat{\mathcal{M}}_{\mathbf{e}_p}\begin{pmatrix} a_1 \\ a_2 \\ a_3 \end{pmatrix} = \begin{pmatrix} -a_1 \\ a_2 \\ a_3 \end{pmatrix} \quad \hat{\mathcal{M}}_{\mathbf{e}_p}\mathbf{b} = \begin{pmatrix} b_1 \\ -b_2 \\ -b_3 \end{pmatrix} \quad (8.27)$$

Here, $\mathbf{a}$ is a vector (like $\mathbf{r}$) and $\mathbf{b}$ is a pseudovector (like $\mathbf{H}$), and without loss of generality it is assumed that the direction of the a_1 component of $\mathbf{a}$ corresponds to $\mathbf{e}_p$.

Trivially, the mirror operator $\hat{\mathcal{M}}$ has the eigenvalues $\lambda_{\pm} = \pm 1$ as $\hat{\mathcal{M}} = \hat{\mathcal{M}}^{-1}$. In the case of an inversion symmetric problem it makes no difference when the inversion is applied before and after $\hat{\Theta}$, $\hat{\mathcal{M}}\hat{\Theta}\hat{\mathcal{M}} = \hat{\Theta}$. In other words $\hat{\mathcal{M}}$ commutes with $\hat{\Theta}$. Hence both operators have a common set of eigenvectors and all modes $\mathbf{H}(\mathbf{r})$ can be classified by their symmetry eigenvalue $\lambda_{\pm}$. Usually modes with $\hat{\mathcal{M}}_{\mathbf{e}_p}\mathbf{H}_{\mathbf{e}_p} = -\mathbf{H}_{\mathbf{e}_p}$ are called $\mathbf{H}$-*even*, and modes with $\hat{\mathcal{M}}_{\mathbf{e}_p}\mathbf{H}_{\mathbf{e}_p} = \mathbf{H}_{\mathbf{e}_p}$ are called $\mathbf{H}$-*odd*. At the mirror plane, i.e., at $\mathbf{r}_{\mathbf{e}_p} = 0$ the even modes must satisfy $-\mathbf{H}_{\mathbf{e}_p}(\mathbf{r}) = \mathbf{H}_{\mathbf{e}_p}(\mathbf{r})$. Hence the $\mathbf{e}_p$ component of the magnetic field of $\mathbf{H}$-even modes vanishes at the mirror plane, and likewise the in-plane component of $\mathbf{H}$-odd modes vanishes.

8.5.2 Translation Symmetries

A dielectric configuration given by $\epsilon_r(\mathbf{r})$ has a translational symmetry when it is unchanged by a displacement by $\mathbf{d}$, or more formally if

$$\epsilon_r(\mathbf{r}) = \hat{T}_{\mathbf{d}}\epsilon_r(\mathbf{r}), \quad (8.28)$$

with the *translation operator* $\hat{T}_{\mathbf{d}}$ defined by

$$\hat{T}_{\mathbf{d}}\mathbf{f}(\mathbf{r}) = \mathbf{f}(\mathbf{r} - \mathbf{d}). \quad (8.29)$$

Since the dielectric configuration is symmetric with respect to the translation operator $\hat{T}_{\mathbf{d}}$, this operator commutes with $\hat{\Theta}$, and both have a common set of eigenvectors. Thus it is convenient to classify the eigenmodes of Θ according to their behavior under translations, which is done in the following.

Discrete Translation Symmetry If Eq. 8.28 is fulfilled for $\mathbf{d} = n\,\mathbf{e}_d$ with integer n, this is called *discrete translation symmetry* in direction $\mathbf{e}_d$. Figure 8.1(a) shows an example of a geometry with

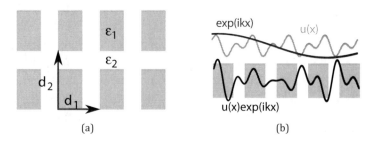

Figure 8.1 (a) Illustration of a configuration with symmetry under translation by $\mathbf{d}_1$ and $\mathbf{d}_2$. (b) Illustration of a Bloch wave Eq. 8.30, formed by the lattice periodic function $u(x)$ and a plane wave $\exp(ikx)$.

such a discrete translation symmetry. The eigenfunctions of $\hat{T}_\mathbf{d}$ are of the form

$$\mathbf{H}'_{m,\mathbf{k}}(\mathbf{r}) = \mathbf{u}_{m,\mathbf{k}}(\mathbf{x})e^{i\mathbf{k}\cdot\mathbf{x}}, \tag{8.30}$$

with eigenvalue $\lambda_\mathbf{d} = e^{-i\mathbf{k}\cdot\mathbf{d}}$ and a periodic function $\mathbf{u}_{m,\mathbf{k}}(\mathbf{x})$ with $\hat{T}_\mathbf{d}\mathbf{u}_{m,\mathbf{k}}(\mathbf{x}) = \mathbf{u}_{m,\mathbf{k}}(\mathbf{x})$. This is the famous *Bloch theorem* known from solid state physics [200, 201]. Now, it is convenient to classify the eigenmodes of Θ by their behavior under translations, i.e., the wave vector $\mathbf{k}$. Here one has to take care that not all values of $\mathbf{k}$ lead to different eigenvalues and eigenfunctions. In particular $\mathbf{k}' = \mathbf{k} + 2n\pi\,\mathbf{d}/|\mathbf{d}|^2$, with any integer number n has the same eigenvalue $\lambda_\mathbf{d}$. Thus for each eigenvalue $\lambda_\mathbf{d}$, represented by $\mathbf{k}$ several modes with different Bloch function $\mathbf{u}_{m,\mathbf{k}}(\mathbf{x})$ and different frequency eigenvalue in Θ exist.

Band Structure As the operator Θ is linear, a small variation by $\delta\mathbf{k}$ results in a small variation in the corresponding Bloch function $\delta\mathbf{u}_{m,\mathbf{k}}(\mathbf{x})$ and frequency eigenvalue by $\delta\omega_m$. Thus, the functions $\omega_m(\mathbf{k})$ are continuous in $\mathbf{k}$. Furthermore, the $\mathbf{H}'_{m,\mathbf{k}}(\mathbf{r})$ and thus $\mathbf{u}_{m,\mathbf{k}}(\mathbf{r})$ must be orthogonal for different m. Hence, at fixed $\mathbf{k}$ the frequency eigenvalues for different m must significantly differ. This is the origin of the photonic band structures discussed later on, in Chapter 12.

Continuous Translation Symmetry If Eq. 8.28 is fulfilled for any $\mathbf{d} = a\,\mathbf{e}$ with arbitrary real a the system has a *continuous translation symmetry* in direction $\mathbf{e}$. In this case all $\mathbf{k}$ differ at least in the eigenvalue of one $\hat{T}_{\mathbf{d}}$. Hence $\mathbf{k}$ and the polarization vector ϵ_s completely specify the modes, which have the form of plane waves

$$\mathbf{H}'_{\mathbf{k},s}(\mathbf{r}) = \epsilon_s e^{i\mathbf{k}\cdot\mathbf{x}}. \tag{8.31}$$

This was implicitly assumed in Part I and is the deeper reason for the particular choice of the normal mode field according to Eq. 2.2.

8.6 Total Internal Reflection

Here a situation is considered where one-half space is filled with a dielectric with ε_1, while the other half is filled with ε_2 as illustrated in Fig. 8.2.

Snell's Law Far away from the interface the argumentation from Section 8.3 holds and on both sides the field modes are plane waves satisfying the dispersion relation Eq. 8.20. As the structure is invariant under translations parallel to the interface, a common parallel component of the $\mathbf{k}$-vectors ($\mathbf{k}_{\parallel}$) can be assigned to all field modes. Furthermore, one field mode must have a unique frequency ω_l. From this and Eq. 8.20 one can conclude

$$\frac{\left|\mathbf{k}_{\parallel} + \mathbf{k}_{\perp}^1\right|}{n_1} = \frac{\left|\mathbf{k}_{\parallel} + \mathbf{k}_{\perp}^2\right|}{n_2}, \tag{8.32}$$

or in terms of the angles θ between the different wave vectors defined according to Fig. 8.2:

$$\frac{\sin\theta_1}{n_2} = \frac{\sin\theta_2}{n_1}. \tag{8.33}$$

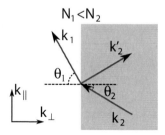

Figure 8.2 Illustration of Snell's law. An incoming wave with wave vector $\mathbf{k}_2$ and angle of incidence θ_2 can be reflected or refracted. Reflection leads to the wave vector $\mathbf{k}'_2$, while refraction leads to $\mathbf{k}_1$.

This equation is commonly known as *Snell's law* and in general has two solutions. The first solution of is that a wave coming from the region with n_2 can be *refracted* into the region with n_1, while the perpendicular component of the wave vector changes accordingly. The second solution is simple reflection, i.e., the wave is reflected at the boundary and one finds with $n_1 = n_2$ the well-known rule that the angle of incidence equals the angle of reflection.

A more rigorous treatment giving the ratio between reflected and refracted intensity can be found in Ref. [12].

Total Internal Reflection From the increased density of states in dielectrics compared to vacuum (cf. Section 8.3), one might guess that there exist some modes that are completely confined within the dielectric. Using Snell's law Eq. 8.33 this can be further investigated, leading to the phenomenon of *total internal reflection*.

Without loss of generality it is assumed that $n_1 < n_2$. In the case that $\sin \theta_2 > n_1/n_2$, i.e., for large angles of incidence, Eq. 8.33 requires $\sin \theta_1 > 1$. This is clearly impossible with a real wave vector $\mathbf{k}_1$. Hence such waves cannot propagate in the region with lower refractive index and the wave is totally reflected at the interface when its angle of incidence θ_2 exceeds the critical angle θ_c:

$$\theta_c = \arcsin \frac{n_1}{n_2}. \tag{8.34}$$

Although there are no propagating waves within the low index medium, the evanescent field in the vicinity of the surface might become very strong.

In the usual configuration, this effect confines light within a dielectric. Hence this phenomenon is called total internal reflection and is a major hurdle for light extraction from semiconductor devices [63] as it overcompensates the gain in photon emission according to Eq. 8.21. A conceptually simple method to overcome this problem is the use of immersion microscopy. This will be introduced in the next chapter, whereas applications of total internal reflection for light confinement are presented later on.

Chapter 9

Immersion Microscopy

Immersion microscopy is a general term for techniques to modify the geometry of the dielectric in a way that emission from at least one point is almost normal incident. By this, total internal reflection can be prevented. Hence immersion microscopy is ideal to profit from the enhanced fluorescence intensity of point-like quantum emitters embedded in dielectrics.

Mainly two immersion techniques that are suitable to observe single quantum emitters can be distinct: liquid immersion microscopy and solid immersion microscopy, both discussed in what follows.

9.1 Liquid Immersion Microscopy

Liquid immersion microscopy is the oldest immersion technique. Here the dielectric matrix of the quantum system is optically connected to the objective lens by an index matched immersion medium. In this ideal case depicted in Fig. 9.1, refraction occurs only at the curved surface of the objective lens. Thereby total internal reflection can be prevented and the enhanced spontaneous emission

Integrated Quantum Hybrid Systems
Janik Wolters
Copyright © 2015 Pan Stanford Publishing Pte. Ltd.
ISBN 978-981-4463-82-9 (Hardcover), 978-981-4463-83-6 (eBook)
www.panstanford.com

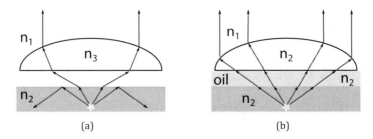

(a)   (b)

Figure 9.1 Conventional and oil immersion microscopy. (a) Illustration of conventional microscopy. A point-like emitter is embedded into dielectric with refractive index n_2. Parts of the emission undergo total internal reflection and cannot be collected by the objective lens. (b) Immersion microscopy overcomes this problem by placing an immersion medium (usually oil or water) which is index matched in between sample and lens.

results in an increased detected intensity when a single quantum emitter is observed.

The numerical aperture

$$NA = n\sin\theta \qquad (9.1)$$

with θ being the half angle of the maximum light cone collected by the lens can reach values of $NA = 1.56$ in oil immersion systems, while in conventional systems $NA > 0.95$ is hardly reached.

Owing to the enhanced numerical aperture the maximal resolution of the microscope as given by the *Rayleigh criterion*

$$R \approx \frac{0.61 \cdot \lambda}{NA} \qquad (9.2)$$

can also exceed conventional systems. This can be understood in terms of the increased number of **k**-vectors contributing to the final image.

Although these advantages are striking, the liquid immersion technique is technologically complicated and neither suitable for efficient light sources like LEDs nor for cryogenic applications. This disadvantage led to the invention of solid immersion lenses.

9.2 Solid Immersion Microscopy

The problem of beating the total internal reflection for efficient light generation came up with the development of light-emitting

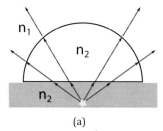 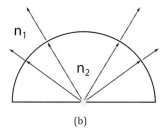

(a)   (b)

Figure 9.2 (a) Illustration of a solid immersion lens with n_2 placed on an index-matched substrate. Similar to liquid immersion microscopy, total internal reflection is inhibited, allowing for enhanced collection efficiency. (b) Illustration of a solid immersion lens with an emitter placed directly on the flat side of the immersion lens. In this configuration the emission profits from the enhanced density of states in the evanescent field of the dielectric.

diodes [202, 203]. The general idea is to get rid of the immersion medium and place a spherical *solid immersion lens* (SIL) directly atop of the substrate as depicted in Fig. 9.2(a). To provide maximum efficiency, the lens can be produced directly into the substrate material. Sometimes it is not even necessary to have a substrate. When a small emitter is placed directly on the flat side of the SIL it can profit from the increased density of states in the high refractive index medium by coupling to the evanescent field [Fig. 9.2(b)]. Here an exact calculation of the spontaneous emission enhancement requires some nontrivial calculations. Analytical solutions for a dipole can be found in Ref. [204]. As a rule of thumb one can assume that the emission into the high-index half-space is enhanced by more than n. While the use of such SILs for enhanced collection efficiency dates back to the 1950s [202, 203, 205], it was not until the early 1990s that an enhanced spatial resolution by solid immersion microscopy was demonstrated [206], and *solid immersion microscope* became its name. There are technological difficulties in shaping the lenses. Nevertheless, today solid immersion lenses are widely used in many experiments for enhancing spatial resolution and collection efficiency [207–209].

Chapter 10

Index Guiding Structures

So far, total internal reflection was seen as a hurdle in efficient light extraction. Nevertheless, it can also be used to confine light within small spatial regions as required for cavity quantum electrodynamics. Following Ref. [199], first the fundamental example of guided modes in a slab is introduced, which serves later on as the foundation of more complex structures like waveguides and disk resonators.

10.1 Guided Modes in Infinite Dielectric Slabs

Infinite dielectric slabs, suspended in vacuum, support at least one guided mode that is bound to the slab. In the following sections these slab modes are first discussed by symmetry consideration, whereas numerical solutions of the master equation follow in Section 10.1.2.

10.1.1 Symmetry Considerations

In the vicinity of an infinite slab with thickness a in z direction [Fig. 10.1(a)] the solutions of Eq. 8.23 cannot be obtained purely by symmetry considerations. Nevertheless, because of the translational

Integrated Quantum Hybrid Systems
Janik Wolters
Copyright © 2015 Pan Stanford Publishing Pte. Ltd.
ISBN 978-981-4463-82-9 (Hardcover), 978-981-4463-83-6 (eBook)
www.panstanford.com

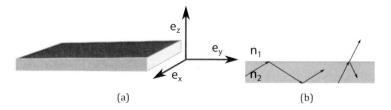

(a)   (b)

Figure 10.1 (a) Sketch of a quasi-two-dimensional plane of dielectric. It is assumed that it expands in x and y directions over the entire space, while its thickness in z direction is limited to t. (b) Illustration of confinement by multiple total internal reflection. While waves with a small $k_z/|\mathbf{k}|$ contribution are confined, others have a steeper angle of incidence and escape.

symmetry, an in-plane wave vector $\mathbf{k}_\rho$ can be assigned and the field modes have the form

$$\mathbf{H}'(\mathbf{r}) = \mathbf{H}'_{i,\mathbf{k}_\rho}(\mathbf{z})e^{i\mathbf{k}_\rho \cdot \rho}, \qquad (10.1)$$

with the in-plane position vector $\rho = \mathbf{e}_x\mathbf{r} \cdot \mathbf{e}_x + \mathbf{e}_y\mathbf{r} \cdot \mathbf{e}_y$.

The slab is mirror symmetric on the x–y plane. Hence modes can be classified into odd and even. Recalling that for **H**-odd modes the in plane components of the magnetic field vanish in the mirror plane and hence the field is mainly electric. These modes are called *transverse electric* (TE) modes. Similar, one can argue that the in plane components of the electric field of the **H**-even modes vanishes at $z = 0$ and hence these modes are named *transverse magnetic* (TM) modes.

10.1.2 *Mode Guiding*

If the thickness of the slab is large compared to the wavelength, one can assume that individual plane waves are multiply reflected at the slab surface and therefore can never leave the slab. The picture illustrated in Fig. 10.1(b) is a good approximation of an elaborate analytical theory that can be found in Ref. [210]. If the thickness is small compared to the wavelength the assumption of plane waves in direction of z within the slab cannot be motivated. Nevertheless, one can show that in this case guided modes exist within the slab.

As an exact analytic expression could not be found yet, the master equation Eq. 8.23 or Maxwell's equations are solved numerically.

Several methods exist among which the method of *plane wave expansion* [211] is the most convenient for the problem.

The susceptibility $\varepsilon_r(\mathbf{r})$ and the **H**-field are expanded into plane waves, i.e., into their spatial Fourier components. Then, the problem is discretized and assumed to be periodic, resulting in a finite set of Fourier coefficients. These can be inserted into the master equation giving a finite set of coupled linear equations for each **k** vector, which can be efficiently solved yielding the normal modes and their frequency.

Several open source implementations of the algorithm exist, among which the *MIT photonic bands package* (MPB) is the most popular. Using MPB, the dispersion relation of the modes of the infinite slab can easily be found. Here, one has to take care that not all modes are actually bound to the slab. Modes that fulfill $\omega > c|\mathbf{k}_\rho|$ have traveling components escaping to the surrounding vacuum as discussed in Section 8.6. This results in the so-called *light cone*: within the light cone modes are unbound and only modes with $\omega < c|\mathbf{k}_\rho|$ are bound to the slab.

Figure 10.2 shows the numerically calculated modes of slab waveguides with thickness t and refractive index of $n_{SiN} = 2.0$ and $n_{GaP} = 3.4$ corresponding to Si_3N_4 and GaP, respectively. In both cases, bound TE and TM modes exist, of which only the lowest are

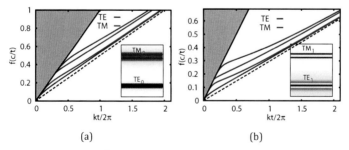

(a)   (b)

Figure 10.2 Numerically calculated dispersion of slab waveguides. (a) Dispersion of the first guided modes in a slab of thickness t with $\epsilon_r = 4$ corresponding to Si_3N_4. The inset show the profile of the electric field intensity at $kt/2\pi = 0.25$ projected into the x–z plane, with darker color indicating higher intensity. The dashed line indicates the dispersion in bulk material of same refractive index. The shaded area corresponds to the light cone of unconfined waves. (b) Same as in (a), but for a slab with $\epsilon_r = 11.6$ corresponding to GaP. At $kt/2\pi = 0.25$ four modes are confined.

shown. While for the higher modes a cut-off frequency exists, below which bound states do not occur, this is not the case for the lowest-order modes (TE_0, TM_0). These are bound even for low frequencies. For high frequencies their dispersion relation is comparable to the corresponding bulk material Eq. 8.20, while the vacuum dispersion is approached in the low frequency limit. Furthermore, a lowering of the cut-off frequency of higher order modes can be observed with increasing optical thickness $n \cdot t$ of the slab. This is in agreement with the assumption of multiply reflected plane waves as illustrated in Fig. 10.1(b). The electric field profile of the TE modes shows the expected field minimum in the symmetry plane, while the TM modes show a field maximum.

10.2 Strip Waveguides and Fibers

While the slab modes show confinement only in one direction, the mode guiding can also be used to achieve two-dimensional confinement by cutting out a strip of the slab. In this case two-dimensional confinement can be achieved, which is very useful for efficient wave guiding on photonic chips [212, 213]. Figure 10.3(a)

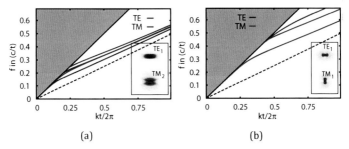

(a)   (b)

Figure 10.3 Numerically calculated dispersion of waveguides. (a) Dispersion of the first guided modes in a strip of thickness t and width $w = 3t$ with $\epsilon_r = 4$ corresponding to Si_3N_4. The inset shows the profile of the electric field intensity at $kt/2\pi = 0.25$ projected into the x–z plane, with the darker color indicating higher intensity. The dashed line indicates the dispersion in bulk material of same refractive index. The shaded area corresponds to the light cone of unconfined waves. (b) Same as in (a), but for cylindrical waveguide with diameter t. The two lowest TE and TM modes are degenerate and frequently used for telecommunication applications. The inset shows the profile of these two modes.

shows the numerically calculated band structure of a free standing waveguide with thickness a and width $w = 3a$ made of Si_3N_4. Remarkably, as only a refractive index contrast between waveguide and environment is necessary, the waveguide can also operate when suspended on a substrate. The same concept of index guiding can also be applied to extended cylinders, e.g., *optical fibers* [Fig. 10.3(b)]. For these, the wide field of applications in telecommunication [214] is probably most prominent. Recently, ultrathin tapered fibers have drawn much research effort, as they allow for comparably strong coupling of emitters to the evanescent field [61, 215, 216].

10.3 Whispering Gallery Modes in Disk Resonators

While waveguides allow optical confinement in two dimensions, it is also possible to obtain three-dimensional index confinement. For this, the *disk resonators* are conceptually the simplest examples. Such a resonator consists of a dielectric disk of diameter d and with thickness t, as illustrated in Fig. 10.4(a).

If the diameter of the resonator is large compared to the wavelength, at a first point the modes can be approximated by the

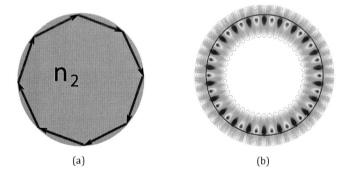

(a)   (b)

Figure 10.4 The whispering gallery mode resonator. (a) Illustration of a whispering gallery mode with $l = 8$. (b) Simulation of the electric field in the x–y plane of a mode with $n = 1$ and $l = m = 22$ of a 2 µm disk with $n = 1.5$ (fused silica). Solid line indicates dimension of the resonator. Image courtesy of R. Henze.

slab modes from section 10.1. Hence there are traveling plane waves inside the disk. Now it might appear that such a wave undergoes repeated total internal reflection at the boundary of the disk, as illustrated in Fig. 10.4(a). In case of many reflections, a wave is resonant with the disk, when its circumference is a multiple integer of the wavelength. Hence the frequency of a mode with l bounces is calculated to be approximately

$$\omega_{res} = \frac{2lc}{dn_{eff}}, \tag{10.2}$$

with the effective refractive index n_{eff} defined by $n = c/c_{eff}$, where c_{eff} is the group velocity of the wave. This gives directly the definition of the angular mode number l, being close to the ratio between circumference of the disk and wavelength [217], i.e., $l \approx \pi d/(n_{eff}\lambda)$. Furthermore the modes can be classified by their number of radial nodes n and the number of azimuthal nodes m, i.e., the number of bounces on the surface. As such modes where first described by Lord Rayleigh to explain the phenomenon of the whispering gallery in the St. Paul's Cathedral in London they are called *whispering gallery modes* (WGM). These are not limited to disks, but can also be observed in any resonator with near circular symmetry like toroids [218], spheres [219] or bottles [220]. Figure 10.4(b) shows an example of the electric field distribution in the x-y-plane of a mode with $n = 1$ and $l = m = 22$ of a 2 μm disk with $n = 1.5$. The quality factor of such structures is usually limited by surface roughness, leading to scattering and thus reduced internal reflectivity [221].

In the following sections, first a simple method for fabrication of WGM resonators is discussed, before a technique for measuring the mode structure is described.

10.3.1 *Fabrication of Disk Resonators*

The fabrication process of disk resonators depends naturally on the material of choice. For silicon-based semiconductors, the hole CMOS manufacturing processes are available. From a silicon wafer with a thin layer of silica (SiO_2) disks can be made by *electron beam lithography* and subsequent removal of the underlying silicon. To lower the surface roughness the outer rim of the disk can be melted using a focused CO_2-laser resulting in a high-quality microtoroid [218]. Today Q factors of $Q > 10^8$ and mode volumes of $V \approx 100$ µm^2 are frequently realized [219, 222].

An alternative route to fabricate microdisk resonators is the use of *direct laser writing* (DLW) in polymer. For this, a femtosecond laser beam is focused into a photoresist. By multiphoton absorption the photoresist polymerizes, not only enabling a sequential three-dimensional exposure by scanning the tightly focused beam over the sample [cf. Fig. 10.5(a)], but also structuring below the standard Rayleigh limit Eq. 9.2 [223, 224]. Using either positive or negative resist the exposed or unexposed parts are resistant to a solvent, while other parts can be removed in the development process.

Apart from the cost efficiency of this process, the ability of fabricating true three-dimensional structures is clearly advantageous compared to the standard CMOS process. So, resonators with Q factors of $Q \approx 10^6$ [225, 226] and multidimensional waveguides [227], as illustrated in Fig. 10.5(b), can be written in one single exposure step [198].

10.3.2 *Measurement of the Mode Structure of Disk Resonators*

Usually, much effort is spent to increase the resonators' Q factor defined in Section 4.4.1 by Eq. 4.104, i.e., to minimize the

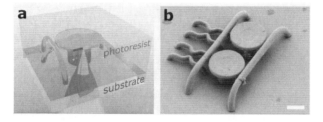

Figure 10.5 (a) Illustration of the direct laser writing process. A photoresist spin-coated onto a transparent substrate, e.g., a cover slip, is exposed via multiphoton polymerization by a tightly focused femtosecond laser. This allows for true three-dimensional fabrication. (b) Example of a direct laser written structure consisting of two WGM disk resonators and several waveguides. The white scale bar is 5 µm. From Ref. [198], © 2013 Nature Publishing Group.

coupling of the resonator modes to free space. Hence free-space in- and out-coupling is cumbersome and a measurement of the cavity resonances requires alternative methods. In lithographic semiconductor microdisks a very common approach is the use of fiber tapers, i.e., thin cylindrical waveguides. These are brought into the near field of the resonator. Here the evanescent field of the fiber taper and the resonator overlap, allowing for efficient in- and out-coupling. This approach requires the accurate positioning of macroscopic fibers with a precision of a few nanometers to control the coupling strength. Such measurements are achievable in a laboratory environment and the narrow cavity modes appear as dips in the transmission spectrum of the fiber (cf. Fig. 10.6).

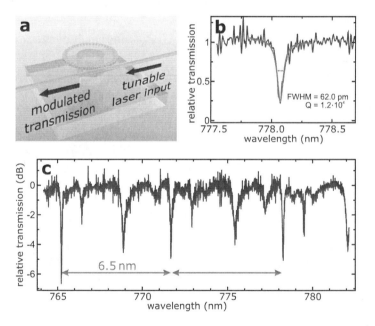

Figure 10.6 (a) Illustration of the measurement scheme. A fiber taper is brought into the near field of the WGM. Owing to coupling into the resonator the transmission drops in resonance. (b) The resonance of a polymer resonator measured with a tunable laser has a width of 62 pm at a wavelength of 778 nm, corresponding to a Q factor of $Q = 1.2 \cdot 10^4$. (c) The spacing between two principal modes is 6.5 nm in accordance with Eq. 10.2. From Ref. [198], © 2013 Nature Publishing Group.

For large-scale integrated applications, coupling of tapered fibers is clearly not applicable. Here, the three-dimensional DLW process has a main advantage. Using DLW the necessary waveguides can directly be fabricated on chip with precise controllable distance and coupling strength as illustrated in Fig. 10.5.

Chapter 11

Photonic Crystals

11.1 Introduction to Photonic Crystals

Photonic crystals are dielectric arrangements with discrete translational symmetries. Depending on the dimensions in which the symmetries exist, one can distinguish between one-, two-, and three-dimensional photonic crystals. According to Section 8.5, the solutions of the master equation are Bloch waves Eq. 8.30, and the eigenfunctions with a specific k vector can be ordered into photonic bands according to the eigenfrequency ω, starting with the lowest eigenfrequency value.

Remarkably, it is possible to construct periodic structures in which a certain frequency range around $f_0 = c/\lambda_0$ is not covered by any band, i.e., a *photonic bandgap* exists. Probably the most simple and intuitive example of a one-dimensional structure with bandgap is the $\lambda_0/4$ stack as illustrated in Fig. 11.1(a). Today, such devices are frequently used as *Bragg mirrors* in bulk, as well as in integrated optics. In detail, the stack consists of alternating layers of thickness $d_i = \lambda_0/(4n_i)$, made of two dielectrics with different refractive indixes n_i. It is invariant under any lateral translation, whereas along the stack axis translations by the *lattice constant* $a = d_1 + d_2$ leave it unchanged. In the structures an incident beam is multiply

Integrated Quantum Hybrid Systems
Janik Wolters
Copyright © 2015 Pan Stanford Publishing Pte. Ltd.
ISBN 978-981-4463-82-9 (Hardcover), 978-981-4463-83-6 (eBook)
www.panstanford.com

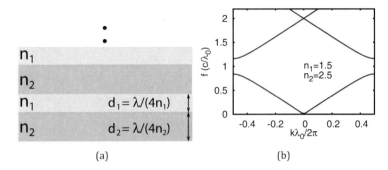

(a) (b)

Figure 11.1 (a) Illustration of the $\lambda/4$-stack. (b) Calculated band structure of a stack with $n_1 = 1.5$ and $n_2 = 2.5$ corresponding to SiO_2 and TiO_2, respectively. As the components parallel to the layers are plane waves, only the perpendicular k vector component is shown. For the designed wavelength λ_0 a bandgap exists and light cannot propagate through the stack.

reflected at the interfaces between the different materials, leading to total destructive interference in the direction of propagation for frequencies close to f_0. This can easily be understood in terms of the numerically calculated band structure shown in Fig. 11.1(b). Around the frequency f_0, i.e., when the wavelength in the dielectric is twice the lattice period, the structure exhibits a bandgap. In this bandgap no solutions of the master equation exist and hence light propagation is forbidden.

Today many more two- and three-dimensional geometries with photonic bandgap are known. Nevertheless, the $\lambda/4$ stack is the only one with a simple analytic solution. Hence it is not surprising that it took about a century until progress in computational power allowed the discovery of further geometries. Here, the most prominent three-dimensional examples are the seminal *Yablonovite* [228], the *inverse opal* [229], and a diamond lattice of air holes [230]. While in nature such photonic crystals are responsible for the colors of butterfly wings [231] and peacock feathers [232], even with today's technology fabrication of such structures is still challenging and only few experimental realizations exist, e.g., Refs. [223, 233–235].

It is also possible to design two-dimensional structures with a partial bandgap for certain mode symmetries. Here, the lattice of dielectric rods in air and the trigonal lattice of air holes in dielectrics are the most prominent examples. While the first one can hardly be realized, the latter one is much simpler to fabricate and frequently used in many experiments. In the following, these two-dimensional photonic crystals are introduced.

11.2 Photonic Crystal Slabs

11.2.1 *Geometry and Band Structure*

The two-dimensional slab with thickness t from Section 10.1, modified by a trigonal lattice of air holes with radius r and lattice constant a as illustrated in Fig. 11.2(a) is a geometry with a partial bandgap [236]. Here, light is multiply scattered at the holes and therefore might destructively interfere. Calculations of the band structure show that the bandgap is neither complete, nor does it exist for all bands. Nevertheless, within the light cone, where modes are bound to the slab, it is possible to find a bandgap for all TE modes, i.e., for those modes confined in the slab with a minimum of the electric field in the slab center (cf. Fig. 10.2). For these TE modes the vacuum fluctuations vanish in the slab and spontaneous emission into slab modes can be suppressed.

The size of the TE bandgap, given by the ratio between gap width and mid-gap frequency, depends strongly on the geometry

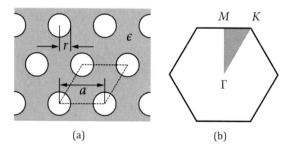

(a)   (b)

Figure 11.2   (a) Geometry of the hexagonal lattice with lattice constant a, hole radius r, thickness t (not shown), and relative permittivity ϵ_r of the dielectric. Dashed lines indicate the primitive lattice cell. (b) First Brillouin zone of the hexagonal lattice. By symmetry considerations, it can be reduced to the shaded area between the symmetry points Γ, M, and K.

parameters. Figure 11.3 shows two example band structures for $r/a = 0.3, t = 0.5a$ with $n = 2$ and $n = 3.4$, corresponding to silicon nitride and gallium phosphide, respectively. While the comparable low-index contrast between dielectric and the holes allows only for a 10% gap in case of Si_3N_4, the larger-index contrast in GaP gives a 26% bandgap. The position of the bandgap center corresponds roughly to $f_{cen} \approx c/(na)$.

To optimize the bandgap size, numerical simulations for different air hole radii r and slab thicknesses t can be performed (Fig. 11.3). While in principle the gap size increases with growing radius, this stops when the wings of the second band get into the light cone, resulting in a shrinking gap. This is in particular an issue for larger slab thicknesses t, when the gap might be completely avoided.

Even more important for the design of photonic crystal structures for applications is the width of the bandgap at its low frequency edge. This gap width, measured in fractions of the path Γ-M-K-Γ, increases with shrinking hole size and growing slab thickness. Thus a thicker slab with $t \sim a$ with small holes $r \sim 0.25a$ is in general desirable. Nevertheless, such slabs are hard to manufacture, and hence the production process usually gives a lower limit for the hole size.

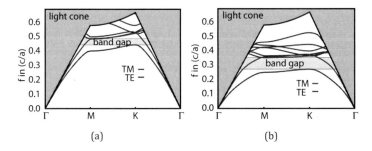

(a) (b)

Figure 11.3 Band structures of photonic crystal slabs with $r/a = 0.3$, $t = 0.5a$. (a) First TE and TM modes of a photonic crystal in silicon nitride with $n = 2.0$. Within the light cone, where modes are bound to the slab, exists a gap for TE modes between $f = 0.44$ c/a and $f = 0.49$ c/a, corresponding to a gap mid-gap ratio of 9.6%. (b) Same as (a) for gallium phosphide with $n = 3.4$. Here the gap is between $f = 0.27$ c/a and $f = 0.35$ c/a, corresponding to a much larger bandgap of 26%.

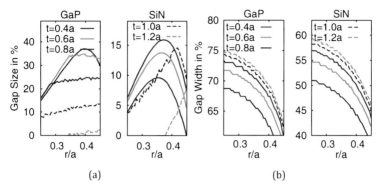

Figure 11.4 Optimization of the bandgap for Si$_3$N$_4$ and GaP. (a) Size of the bandgap given by the gap mid-gap ratio as function of hole radius for various slab thicknesses between $t = 0.4a$ and $t = 1.2a$. (b) Relative width of the photonic bandgap, where the full path Γ-M-K-Γ is defined as 100%.

11.2.2 *Fabrication*

The two-dimensional structure of the photonic crystal slabs allows fabrication by lithography and etching, comparable to the CMOS process. The required fabrication precision can be estimated from the lattice constant a: Photonic crystals matching the spectral properties of solid quantum emitters must provide a photonic bandgap in the visible to near infrared, i.e., around 640 nm for NV centers or up to about 900 nm for InAs quantum dots in GaAs. As the bandgap center frequency is always near $f_c \approx c/(na)$, this requires the lattice constants to be on the order of λ/n, with the size of the holes being a fraction of that. Hence, the required resolution of the production process on the order of < 100 nm. This is hardly

Table 11.1 Materials for photonic crystals

material	refractive index	direct bandgap	Ref.
Si$_3$N$_4$	2.0	5.4 eV	[237]
GaP	3.3	2.9 eV	[238, 239]
GaAs	3.6	1.4 eV	[239, 240]
C (diamond)	2.4	5.5 eV	[240]

Typical materials for photonic crystals matching solid quantum emitters. All values are given at 300 K.

achievable with conventional optical lithography. Here electron beam lithography, providing the resolution of standard scanning electron microscopes has proven to be a suitable technique. As membrane material any dielectric being transparent in the desired wavelength range can be used. Table 11.1 gives an overview over common photonic crystal material systems. For NV centers Si_3N_4 and GaP are particular suitable. Both materials differ slightly in the fabrication process, on which the following paragraphs focus.

Fabrication of Photonic Crystals in Si_4N_4 Fabrication starts with commercial low-pressure chemical vapor deposition (LPCVD) grown Si_3N_4 layers on a sacrificial silica (SiO_2) layer on silicon (Si). Typically the thickness of the silicon slab is 525 µm, the thermally oxidized SiO_2 layer has a thickness of 2.2 µm, while the Si_3N_4 layer has a thickness of 230 nm. By spin-coating a polymethylmethacrylate (PMMA) positive resist is deposited on the wafer and dried in a vacuum oven (cf. Fig. 11.5, Step I).

The pattern is written by means of a 30–100 kV scanning electron microscope using a suitable lithography unit. Afterwards the exposed resist is developed and the development process

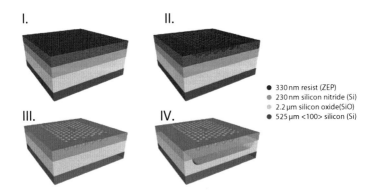

I. II. III. IV.

● 330 nm resist (ZEP)
● 230 nm silicon nitride (Si)
● 2.2 µm silicon oxide(SiO)
● 525 µm <100> silicon (Si)

Figure 11.5 Fabrication of photonic crystal slabs. Step I: The wafer is covered with suitable resist material. Step II: Using electron beam lithography, the designed structure is written into the resist and developed. Step III: The mask is transferred into the membrane by reactive ion etching. Step IV: In a last step the substrate below the membrane is removed by selective etching. Image courtesy of M. Schoengen.

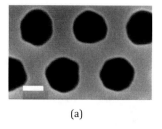
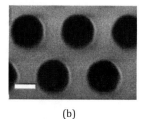

(a)   (b)

Figure 11.6 Examples of fabricated photonic crystals in Si_3N_4. Scale bar is 100 nm. In (a) the etching parameters are not well adjusted, resulting in disordered air holes. In (b) the parameters are better adjusted, leading to circular shaped well-ordered holes. Image courtesy of M. Schoengen, J. Probst.

stopped by a stopper. Subsequent rinsing in isopropyl alcohol and deionized water exposes the written pattern (cf. Fig. 11.5, Step II).

Now, the pattern can be transferred onto the silicon nitride layer by reactive ion etching with octafluor cyclobutane (C_4F) and sulfur hexafluoride (SF_6), e.g., in a plasma etcher. The etching has to be fully anisotropic to have little lateral etching. Here, reactive gas composition, pressure, and flow as well as power density of the plasma are important and sensitive parameters that require fine adjustment to produce well-shaped holes. (cf. Fig. 11.6).

In a last step the sacrificial SiO_2 layer below the Si_3N_4 is dissolved. This is done by selective wet etching using 5% of buffered HF (50% H_2O with 50% HF) in ammonium fluoride (NH_4F). As opposed to dry etching, wet etching is isotropic, i.e., the etching rate is the same in all directions [241]. Here, one has to take care that the selectivity is not perfect and the Si_3N_4 layer thickness is reduced, while the hole radius gets slightly increased. After this step, the suspended photonic crystal slabs have been produced, cf. Fig. 11.5, Step IV.

Fabrication of Photonic Crystals in GaP   The fabrication steps of photonic crystals in GaP are in principle the same as for Si_3N_4 structures described in the previous paragraph. Nevertheless, here the wafer material is not commercially available and thus the process has to rely on individually grown GaP layers. For example, thin gallium phosphide (GaP) layers, e.g., 60 nm can be deposited

Figure 11.7 Example of photonic crystal in GaP. To allow for expansion after fabrication, the photonic crystal membrane is mounted in spring-like supports. Image courtesy of M. Schoengen.

on a silicon (100) substrate using heteroexpitaxy [242–244]. In this case, the sacrificial layer is missing and the bulk silicon takes this task. The lithography step equals the previously described one, where other resists might be used (Steps I–II). Again, the pattern transfer to the GaP layer (Step III) can be performed by dry etching, where using boron trichloride (BCl_3) in this case. The subsequent removal of the underlying Si sacrificial layer (Step IV) might be done by isotropic dry etching with sulfur hexafluoride (SF_6). To remove remaining resist, the hole sample is cleaned in an oxygen plasma.

Importantly, due to a tiny lattice mismatch, the GaP layer is strained and may bend when it is under etched. To allow for expansion without bending, spring-like membrane supports (see Fig. 11.7) can be used.

11.3 Photonic Crystal Waveguides

The major functionalities of photonic crystal slabs arise from breaking the translational symmetry. For example, when connecting an unpatterned slab with a photonic crystal, bound slab modes inside the bandgap are reflected when impinging on the photonic crystal. The photonic crystal acts as a Bragg mirror for these bound modes.

This concept can be further expanded to photonic crystal waveguides, where a tiny quasi-one-dimensional slab region of width W is embedded between two photonic crystal slabs as

Figure 11.8 Scheme of a W1 photonic crystal waveguide with width $W = 1 \cdot \sqrt{3}a$, created by removal of a row of air holes in an otherwise perfectly ordered photonic crystal slab.

illustrated in Fig. 11.8. In case of a W1 waveguide, where $W = \sqrt{3}a$, this can also be understood as removing a row of holes in the crystal. Thereby its symmetry is broken [245, 246], and as known from lattice defects in semiconductors, this may give rise to additional states within the bandgap as illustrated in Fig. 11.9. These states are index confined in the z direction, i.e., perpendicular to the slab, while bandgap confined in the y direction. Hence propagation can only occur in the x direction, along the waveguide. The bandgap confinement has important advantages above the pure index-confined waveguide as introduced in Section 10.2. While the index confinement by total internal reflection gives a strict limit on the

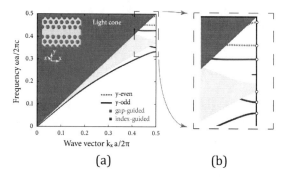

(a)   (b)

Figure 11.9 (a) Projected band structure of the W1 waveguide with $r = 0.375a$, $t = 0.8a$ and $n = 2$. The light shaded area corresponds to the continuum of photonic crystal slab modes that are not confined in the waveguide. Modes below this continuum are index confined, while modes above the continuum are bandgap confined. (b) Zoom into the relevant upper right corner of (a). Around $f \approx 0.46$ the waveguide is single mode with a flat dispersion allowing for slow light.

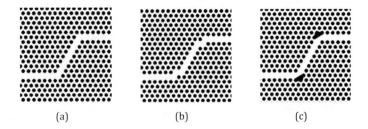

(a) (b) (c)

Figure 11.10 Illustration of photonic crystal waveguide bends. (a) Unoptimized 60° bend. (b) Optimization of the bend by displacing one hole at the bend. (c) Further optimization can be achieved by introducing elongated slits at the bend.

minimal bending radius, this is not the case for bandgap-confined modes. Here, high transmission can be reached even for sharp bends [247]. As illustrated in Fig. 11.10, in the recent years several designs have been developed to further improve this property [248–250] and to allow for high-density integration of photonic circuits using photonic crystals.

When changing the width of the W1 waveguide by a factor of x a Wx waveguide is generated. In this waveguide the guided modes are slightly shifted to higher frequencies for $x < 1$, while they are downshifted for $x > 1$. If the width is properly altered, the waveguide becomes single mode at a desired frequency [251].

11.4 Photonic Crystal Cavities

While a line defect in the photonic crystal results in two-dimensional confined states within the bandgap, it is also possible to achieve full three-dimensional confinement with point-like defects. This results in photonic crystal cavities, providing ultrasmall mode volumes on the order of the cubic wavelength and reasonable high Q factors. Two approaches can be distinguished: the point-like modulation of an otherwise perfect photonic crystal and the point-like modulation of a photonic crystal waveguide. The most prominent representative of the first category, the *L3 cavity* is introduced in the following paragraph, while *modulated waveguide cavities* are described later on in Section 11.4.3.

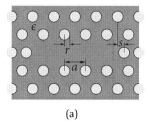
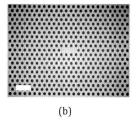

(a)   (b)

Figure 11.11 (a) Scheme of the L3 photonic crystal cavity created by removal of a row of three air holes in an otherwise perfectly ordered photonic crystal slab. For optimization of the cavity, the outer holes might be displaced by s (cf. Section 11.4.2). (b) Scanning electron micrograph of an L3 photonic crystal cavity realized in Si_3N_4. Scale bar is 1 μm.

11.4.1 *L3 Cavity*

The *L3 cavity* consists of a row of three missing holes in the trigonal photonic crystal lattice as illustrated in Fig. 11.11. This lattice defect introduces several bound defect states, of which the first is situated at the low frequency edge of the bandgap.

To investigate the properties of the L3 cavity the *finite-difference time-domain method* (FDTD) [253] is suitable. The space around the photonic crystal cavity is discretized, while the partial derivatives in Maxwell's equations are transformed into finite differences. The remaining finite difference equations for all electric and magnetic field components can be solved in a leapfrog manner, allowing a fast and easy calculation of the field evolution. Here, the cavity resonances can be found as a quasi-stationary solution, when placing absorbing boundary conditions. For such simulations, numerous commercial (RSoft, Lumerical FDTD Solutions) and open-source packages (Meep [254]) exist.

Figure 11.12(a) shows the calculated mode profile of the first mode of the L3 cavity. A discussion of higher order modes can be found, e.g., in Ref. [255]. The mode is well confined within few lattice periods and has a clear maximum in the cavity center. This gives a mode volume on the order of $V_{eff} \approx (\lambda_{res}/n)^3$. The frequency and quality factor of the mode can be estimated using the method of harmonic inversion [256]. Figure 11.13 shows the calculated resonance wavelength $\lambda_{res} = c/f_{res}$ and Q factors for a typical

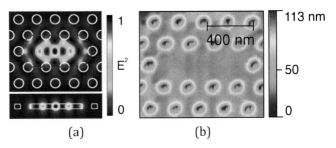

(a)   (b)

Figure 11.12 (a) Simulated (FDTD) electric field profile in the slab center and cross section of the fundamental mode of the cavity. The dashed red line indicates the seven unit cells in which the field is concentrated. (b) Atomic force micrograph of the L3 cavity in GaP. The lattice constant is 209 nm. From Ref. [252], © 2013 IOP.

L3 cavity configurations in GaP ($n = 3.4$). As one might expect from the linearity of the Maxwell's equations, an increase of the lattice constant a goes in hand with linear increase of the resonance wavelength [Fig. 11.13(a)]. Remarkably, a similar linear relation holds also for variation of the air hole radius r [Fig. 11.13(b)]. This can be understood in terms of an effective average refractive index. When the air hole size is increased, the effective index of refraction decreases, which is analogous to shrinking of the entire structure. The same holds also for a direct variation of the refractive index (Fig. 11.14). Remarkably also the Q factor decreases with increasing radius r, i.e., when the bandgap gets narrower. This counterintuitive

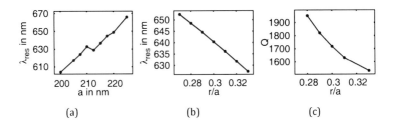

(a)   (b)   (c)

Figure 11.13 Simulation of the fundamental L3 resonance for a GaP slab ($n = 3.4$) with thickness $t = 59$ nm. (a) Dependency of the resonance wavelength λ_{res} on the lattice constant for fixed $r/a = 0.29$. (b) Resonance wavelength λ_{res} for varying radius at fixed $a = 217.5$ nm. (c) Behavior of the Q factor for same variation as in (b). Lines are guides to the eyes.

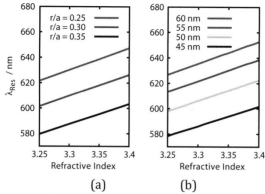

Figure 11.14 (a) Simulated dependency of the first cavity resonance wavelength of a L3 cavity with $a = 209$ nm on the refractive index n for various hole radii r/a (a) at fixed $t = 55$ nm. (b) Same for varying slab thicknesses t at fixed $r/a = 0.3$. In both cases the behavior is nearly linear and in the studied regime the slope does not depend on the thickness of the slab or on the hole diameter. From Ref. [252], © 2013 IOP.

behavior is discussed in the next section, where an optimization strategy for the quality factor is discussed.

11.4.2 *Optimized L3 Cavity*

If the dielectric material is lossless, the cavity Q factor is limited only by scattering from the cavity. The total loss rate $\kappa_{ideal} = \omega/Q_{ideal}$ of an ideally fabricated photonic crystal cavity can be decomposed into losses into the membrane $\kappa_{||} = \omega/Q_{||}$ and losses into the surrounding air $\kappa_\perp = \omega/Q_\perp$:

$$\frac{1}{Q_{ideal}} = \frac{1}{Q_{||}} + \frac{1}{Q_\perp}. \tag{11.1}$$

If the perfect photonic crystal surrounding the cavity is large enough, losses into the membrane can be neglected ($1/Q_{||} \approx 0$) and the quality factor is limited by out-of-plane losses arising from the incomplete nature of the bandgap.

Applying a spatial Fourier transformation to the mode profile of the cavity resonance, one finds that many different in-plane k components contribute [257, 258]. In particular, small k vectors that lie within the light cone and hence couple to free space modes are present. In fact, these components are responsible for the out-of-plane losses. Hence the losses can be reduced and the Q factor increased when these lossy k components get suppressed by a suitable geometry modification. Here one can use the fact that a loser confinement in real space is accompanied by a stronger confinement in Fourier space.

The simplest modification is the displacement of the outer cavity holes by the distance s as indicated in Fig. 11.11 [260]. Figures 11.15 and 11.16 show the calculated influence of the displacement on the Q factor and cavity resonance for Si_3N_4 and GaP photonic crystal slabs. As expected from the analogy to a Fabry–Perot cavity model, the shift leads to a larger extension of the cavity mode,

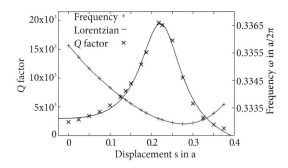

Figure 11.15 Behavior of the fundamental L3 resonance when the outer holes are displace by s. Calculations assume a GaP slab with $r = t = a/4$. As predicted by Ref. [259], the Q factor follows a Lorentzian.

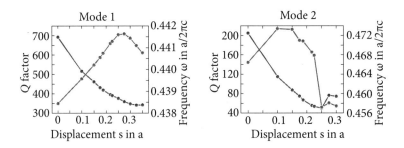

Figure 11.16 Same as Fig. 11.15, but for a Si_3N_4 slab with $r = 0.375a$, $t = 0.8a$. The achievable Q factor of the second order mode 2 is much lower and reaches its optimum at different displacement than the first order mode 1.

resulting in a modified Q factor and resonance wavelength. A more detailed calculation [259] predicts a Lorentzian-shaped relation between Q factor and displacement s, which is well reproduced by the simulations. For GaP, the maximal Q factor is on the order of $Q = 2 \cdot 10^4$, while the Q factor is limited to $Q \approx 10^3$ in SiN. The strong difference can be explained by the different sizes of the bandgap. In the high-index GaP ($n = 3.4$) the bandgap occurs at low frequencies and hence only few k vectors exist outside the light cone (cf. Fig. 11.3). In contrast in Si_3N_4 ($n = 2.0$) with a much lower index of refraction the bandgap occurs at higher frequencies, where the light cone expands widely. Consequently, in this material it is more difficult to construct the cavity mode mainly with wave vectors outside the light cone.

This mechanism also leads to a lower Q factor for higher order modes as visible on the second mode, as illustrated in Fig. 11.16. Furthermore, it explains why the Q factor is increased when the hole size is reduced as illustrated in Fig. 11.13(c). The reduced hole size increases the average refractive index and thereby lowers the edge of the bandgap.

To further improve the quality factor, the upper and lower holes might be shifted and the radius of the holes surrounding the cavity might be reduced, which promises Q factors of up to $5 \cdot 10^3$ even in low-index materials [261, 262]. Nevertheless, in particular the reduction of the holes size is technologically challenging. This impediment is avoided by the high-Q modulated waveguide cavities introduced in the next section.

11.4.3 *Modulated Waveguide Cavities*

By locally increasing the width of a photonic crystal waveguide as illustrated in Fig. 11.17, it is possible to generate localized states. This is in very close analogy to the type I heterostructure in semiconductors from Chapter 5 and hence the resulting photonic cavity is also called *heterostructure cavity*. Here the gradual shift of three rows of air holes allows for smoothly confined modes, promising high Q factors, while the mode volume is only slightly increased to about 6 times the value of conventional L3 photonic crystal cavities. For heterostructure cavities, ultrahigh Q factors of

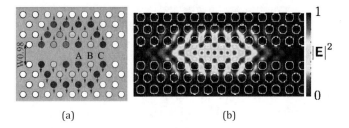

(a)   (b)

Figure 11.17 (a) Illustration of a nanocavity realized by local width modulation of a line defect of width $W = 0.98\sqrt{3}a$. The holes in groups A, B, and C are shifted by distances d_A, d_B, and d_C, respectively, following the arrow to form a tapered shift structure if $d_A > d_B > d_C$. (b) Calculated mode profile of the cavity. The hole shift is too small to be visible.

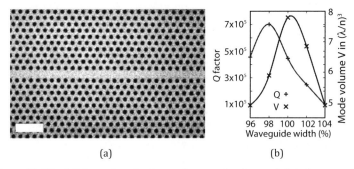

(a)   (b)

Figure 11.18 (a) Scanning electron micrograph of a modulated waveguide cavity in Si_3N_4. The modulation is hardly visible. Scale bar is $1\,\mu m$. (b) Simulated quality factor and mode volume of a line defect with shifted inner holes for different waveguide widths, with $r = 0.26a$ and $t = 0.87a$, $d_A = 0.03a$, $d_B = 0.02a$, and $d_C = 0.01a$ in Si_3N_4 ($n = 2.0$). Solid lines are guides to the eyes.

up to $Q \approx 10^8$ were predicted for silicon ($n = 3.4$) [263–265], and Q factors of $Q \approx 10^6$ could be demonstrated. These are so far the highest Q factors reported in photonic crystal cavities. Although such high Q factors are unachievable with low-index materials like Si_3N_4, simulations of the heterostructure cavity promises Q factors of up to $Q \sim 10^6$ with $n = 2$ (Fig. 11.18).

Apart from the modulation of the waveguide width, cavities can also be created by local modulations of the lattice constant a [266–268], the air hole radius r [269], or the index of refraction via oxidation [270]. Here similar Q factors and mode volumes are achievable.

11.5 Experimental Studies of Photonic Crystal Cavities

To experimentally investigate the mode properties of bare photonic crystal cavities, two far-field methods are available. One can use either the intrinsic material fluorescence or the polarization properties of the cavity modes. The first method is introduced in the next Section 11.5.1, while the latter one is presented in Section 11.5.2. Furthermore, several near-field techniques to map the mode profile [271–273] exist. However, if one is interested in the quality factor and the spectral position of the resonance, the above-mentioned methods are much easier to apply.

11.5.1 *Analysis by Intrinsic Fluorescence*

Because of the existence of lattice defects and impurities, most materials fluoresce at least weakly when exited strongly. This fluorescence occurs even if the excitation wavelength is within the bandgap of the bulk material. In unstructured membranes the fluorescence is rather broadband and flat. In contrast, in photonic crystal cavities this intrinsic fluorescence is modified because of the Purcell effect Eq. 4.120 [261]. This gives rise to a peaked spectral structure, where the position of the peaks corresponds to the cavity resonances, while the width of the peaks is directly linked to the quality factor via its definition Eq. 4.104.

To excite and collect the fluorescence from the structure, a home-built microphotoluminescence setup with a numerical aperture of 0.9, as illustrated in Fig. 11.19, can be used. Here, the large numerical aperture is necessary as the photonic crystal cavities emit mainly under high normal angles into the far field [267], unless geometry modifications are applied [274, 275]. If the material quality is high, the fluorescence signal might be weak and a sensitive high-resolution spectrometer is needed.

Figure 11.20 shows two example spectra of a modulated waveguide and an optimized L3 cavity realized in Si_3N_4 and GaP,

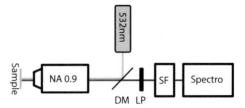

Figure 11.19 Sketch of the optical setup to measure the cavity resonances by using intrinsic fluorescence. Through an objective lens with high numerical aperture (NA 0.9) the photonic crystal cavity is illuminated by a strong laser that is within the bandgap of the dielectric (e.g., 532 nm, 300 µW). Fluorescence light collected through the same objective lens is separated by a dichroic mirror (DM). Residual laser light is filtered by a long-pass filter (LP), while stray light is removed by a spatial filter (SF) consisting of a pinhole in a telescope prior to detection by the spectrometer (Spectro).

respectively. In both cases the achieved Q factor is clearly below the theoretical value. This is mainly due to fabrication imperfections and absorption within the dielectric.

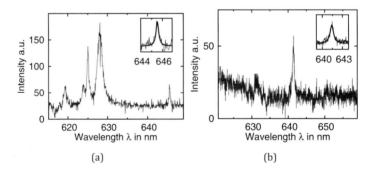

Figure 11.20 (a) Fluorescence spectrum of an optimized modulated waveguide cavity in Si_3N_4, taken under illumination with a green (532 nm) laser at 300 µW power. The first-order cavity resonance appears at 645.5 nm, while higher-order modes are visible below 630 nm. The inset shows a zoom into the first-order mode. The fit unravels a width of 0.3 nm (FWHM), corresponding to $Q = 2150$. (b) Similar experiment with a L3 cavity in GaP. The first-order mode is visible at 641 nm, while higher-order modes appear around 632 nm. The achieved linewidth is 0.6 nm, corresponding to $Q = 1050$.

In high-quality materials, intrinsic fluorescence might be too weak to analyze the cavity modes. In these cases, the photonic crystal can be covered with dye molecules, or other emitters like quantum dots can be incorporated. Nevertheless, this method is not applicable when later on the single quantum emitter shall be coupled to the cavity. In these cases the polarization of the cavity modes helps to get insight into the cavity properties, as discussed in the following.

11.5.2 *Analysis by Polarization Properties*

The far-field emission of the L3 cavity resonance is predominantly linear polarized along the cavity axis [267]. This feature can be used to measure the cavity resonance in the cross polarization scheme [252, 276].

The cavity is illuminated with a vertical polarized collimated broadband beam (white light). For example, a supercontinuum source can be used. This beam is focused on the cavity by a microscope objective with a numerical aperture of 0.9. Importantly, the axis of the cavity is rotated by 45 degrees with respect to the incident polarization. The reflected light is collected through the same objective lens, but only the horizontal polarized component of the reflected light is detected with a 500 mm spectrometer in a confocal configuration. A sketch of the setup is shown in Fig. 11.21.

If the incident light is not in resonance with the cavity, the beam is simply reflected and cannot be detected by the spectrometer. If the light is in resonance with the cavity, there are two effects which slightly rotate the polarization and therefore can be detected on the spectrometer. First, the fraction of incident resonant light which is polarized along the cavity mode (perpendicular to the cavity main axis) couples into the cavity mode. Then the reflected light has a horizontally polarized component which can pass the polarizing beam splitter and reaches the spectrometer. Second, light that was stored in the cavity is reemitted after the characteristic cavity decay time. This light also has a horizontal component. Both components then interfere at the detector and lead to the observation of a Fano-

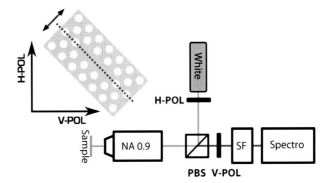

Figure 11.21 Sketch of the optical setup to measure the cavity resonance in the cross-polarized detection scheme. Through an objective lens with high numerical aperture the PC cavity is illuminated with horizontal polarized white light (≈100 nW) from a supercontinuum source. The vertical polarized component of the reflected light is detected in a confocal scheme by a spectrograph. The cavity axis is rotated by 45 degrees with respect to the polarization basis. Thus, the polarization of the reflected light is slightly rotated at the cavity resonance. See text for details. Adapted from Ref. [252], © 2013 IOP.

type resonance of the form:

$$F(\lambda) = A_0 + F_0 \frac{[q + 2(\lambda - \lambda_0)/\Gamma]^2}{1 + [2(\lambda - \lambda_0)/\Gamma]^2}, \tag{11.2}$$

where A_0 is an offset, F_0 is the amplitude of the resonance, λ_0 is the resonance wavelength, $\Gamma = \lambda/Q$ the normalized resonance width, and q the ratio of the amplitudes of the two paths [277]. Figure 11.22 shows an example spectrum of a GaP L3 cavity with lattice constant $a = 209$ nm, a hole radius of 50 nm and the fundamental resonance at about 637 nm, taken at room temperature. The spectrum can be perfectly fitted with three Fano resonances. The fundamental mode supports the highest Q factor of $Q = 580 \pm 50$. With the value from Table 11.1 for the refractive index at $T = 300$ K the slab thickness is estimated according to Fig. 11.14 to be 55 nm.

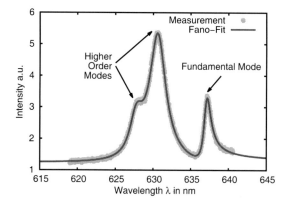

Figure 11.22 Example of the measured spectra and the corresponding fit with three Fano resonances at (637.03 ± 0.01) nm, (630.59 ± 0.01) nm, and (627.68 ± 0.01) nm of an L3 cavity with hole radius $r = 50$ nm. From Ref. [252], © 2013 IOP.

11.6 Tuning of Photonic Crystal Cavities

Because of fabrication tolerances, the resonance wavelength of nominally equal cavities fluctuates by about 1 nm [267]. In order to match the resonances to the narrow emission lines of the quantum emitter, either rigorous selection of suitable cavities or some tuning method has to be applied. For future scalable applications with several cavities, preselection of suitable structures is clearly undesirable, as the yield might be very small when many specific cavities must have equal properties. Thus a cavity-tuning method is clearly preferable. Here, early approaches rely on freezing of gas in cryogenic environment, digital etching [278], or the deposition of small particles in the cavity mode using nanomanipulation techniques [279]. Later on, the tuning by heating was developed. When the cavity material is heated it might oxidize, leading to a decreased refractive index and thus smaller resonance wavelength.

Here, one possibility is to heat the entire sample in an oven [280]. Using this technique all cavities on one chip can be tuned simultaneously in the same way. While this is helpful if one particular cavity shall be tuned to one specific single emitter, it is not suitable if several cavities on the same sample must be tuned to

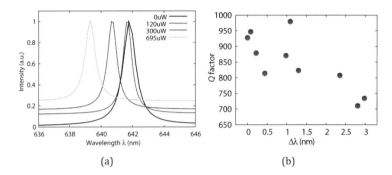

Figure 11.23 (a) Part of the intrinsic fluorescence spectra of a GaP photonic crystal cavity around 640 nm. The fundamental cavity resonance can be permanently tuned by local heating with a focused 407 nm laser. Interestingly, the lineshift depends on the heating power, rather than on the heating time (30 s in all cases). The total tuning range is up to 3 nm. (b) When tuning the cavity resonance, the quality factor remains almost unaffected. Sometimes, even an improvement is possible. Only at larger tuning shifts (about 2 nm), a degradation of the cavity quality factor is measureable. From Ref. [129], © 2013 Wiley-VCH Verlag GmbH & Co.

the same wavelength. Here, a local tuning method like laser-assisted oxidation is required [128, 281].

For example, in GaP a blue (407 nm) laser beam is focused onto an individual photonic crystal membrane. As the photon energy is above the bandgap of GaP, the laser is almost completely absorbed within the membrane and heats the structure locally. It is widely assumed that the material oxidizes [281] and lowers the index of refraction, i.e., the material gets optically thinner. By this procedure the cavity resonances can be irreversibly tuned by up to 4 nm to the blue as illustrated in Fig. 11.23(a). Surprisingly, the wavelength shift does not depend on the heating time but on the heating power. With about 120 µW of heating power the resonance shifts by about 0.1 nm, while it shifts by about 3 nm with 700 µW. At tuning ranges below 2 nm the Q factor remains roughly unaffected (Fig. 11.23(b)). It is even possible to slightly increase the Q factor. Only at highest tuning ranges, the cavity degenerates and the Q factor is significantly lowered. Figure 13.7(c) shows an example of a L3 photonic crystal cavity tuned exactly to the zero photon transition of a nitrogen-vacancy center at 638 nm.

Chapter 12

Applications of Photonic Crystal Cavities

Photonic crystal nanocavities have increasing importance for quantum optics, photonics, and sensing applications [219]. The main reason for this is their ability to confine the electromagnetic field in volumes comparable to the cubic wavelength and thus to strongly enhance photon–matter interaction [116, 260]. This gives rise to a comparably strong interaction between light and single quantum emitters, as will be explained in Part IV. But even in the classical regime, photonic crystals allow for several interesting applications.

The high quality factor of photonic crystal cavities introduced in the previous chapter corresponds to a narrow resonance line. This narrow resonance can be used as optical filter if the cavity is coupled to photonic crystal waveguides, as discussed in the following Section 12.1.

Small shifts of the resonance may occur due to local modifications of the optical properties. Because of the narrow resonance, this can be detected with excellent accuracy. A prominent example of such modifications is the dependency of the index of refraction on the local temperature. In Section 12.2 this is investigated in detail on a photonic crystal cavity from room temperature to 5 K. From this, the index of refraction of the cavity material (GaP) as a basis for a

Integrated Quantum Hybrid Systems
Janik Wolters
Copyright © 2015 Pan Stanford Publishing Pte. Ltd.
ISBN 978-981-4463-82-9 (Hardcover), 978-981-4463-83-6 (eBook)
www.panstanford.com

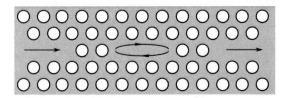

Figure 12.1 Sketch of a photonic crystal filter consisting of a cavity coupled to two waveguides. Light is coupled into the cavity via a waveguide from the left, while the cavity leaks into the waveguide on the right.

predictable design of PC structures in the visible spectral range is derived.

Later on, in Section 12.3 local heating and its effect on the photonic crystal cavity resonances is investigated. This allows experimental studies of thermo-optical switching of visible light. On the basis of theoretical analysis it is estimated that due to the ultrasmall volume of photonic crystal cavities even thermal effects may occur with very short time constants.

12.1 Narrow-Band Optical Filter

When a photonic crystal cavity is coupled to photonic crystal waveguides as illustrated in Fig. 12.1, the cavity mode leaks into the waveguides. This results in a reduced Q factor, according to

$$\frac{1}{Q} = \frac{1}{Q_{cav}} + \frac{1}{2Q_{wg}}, \qquad (12.1)$$

where Q_{cav} is the quality factor of the bare cavity, whereas $1/Q_{wg}$ is proportional to the losses through each waveguide. Although the reduction of Q factor might appear disadvantageous, it gives the possibility to couple light in and out of the cavity on chip. In the device shown in Fig. 12.1, light might be coupled into the cavity from the left, while a detector might be placed on the right. It is intuitively clear that only on resonance light can be coupled into the cavity and hence transmission occurs. In close analogy to a Fabry–Perot etalon the device acts as an *optical filter*, where the transmission T can be defined as the ratio between incoming and transmitted intensity.

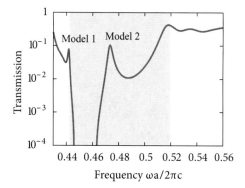

Figure 12.2 Transmission spectrum of a filter built with the unoptimized ($s = 0$) L3 cavity from Fig. 11.16, calculated by FDTD. Within the bandgap (shaded area) two transmission peaks corresponding to the two cavity resonances occur. Owing to parasitary, out-of-plane coupling the transmission does not reach unity on resonance.

Using *coupled mode theory* [199, 282, 283], the maximal achievable transmission can be calculated. For a perfect cavity, where the waveguides are the dominant loss channels, i.e.,

$$Q_{cav} \gg 2Q_{wg}, \tag{12.2}$$

the surprising result is that $T = 1$ on resonance. This can be understood as an incident wave partially couples into the cavity, while most of it gets reflected. On resonance, the intensity inside the cavity will pile up quickly, leading to strong leakage into both waveguides. Now it appears that the light leaking back into the in-coupling waveguide is phase shifted by π compared to the directly reflected wave. Thus both interfere destructively and in a symmetric configuration reflection can be completely inhibited on resonance. In comparison out of resonance the intensity pile-up does not occur and hence most of the light is reflected.

Figure 12.2 shows the calculated transmission spectrum of such a filter. Within the bandgap two transmission peaks occur, corresponding to the two cavity resonances from Fig. 11.16. Nevertheless, the transmission of $T \approx 0.1$ is much less than expected for a perfect system. This can be understood, as Eq. 12.2 is not fulfilled. For the uncoupled cavity the quality factor of the fundamental mode is $Q_{cav} \approx 350$. For the coupled cavity one finds

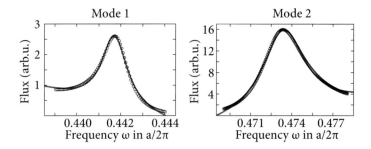

Figure 12.3 Filter transmission line width. Zoom into the transmission peaks from Fig. 12.2. The width of the peaks corresponds to Q factors of $Q_1 = 308$ and $Q_2 = 126$ for the first and second mode, respectively.

$Q \approx 308$ (cf. Fig. 12.3), giving a waveguide coupling of $Q_{wg} \approx 5100$. The cavity is clearly *undercoupled*. For *impedance matching* either the coupling has to be improved [284, 285] or a better cavity design has to be chosen to get high filter transmission.

Such undercoupled filters, where the overall Q factor is only weakly affected by the waveguide, are also frequently applied when measuring the resonances of high-Q cavities [260, 264].

12.2 Refractive Index Measurement in Ultrasmall Volumes

The high quality factor of photonic crystal cavities corresponds to a narrow resonance line. Its shift is a very sensitive indicator for local modifications of the optical properties. A prominent example of such modifications is the dependency of the index of refraction on the local temperature. Cooling of photonic crystal structures is mandatory for many quantum optics experiments, since emission rate and coherence of most quantum emitters improve drastically [286]. However, for complex integrated circuits [116, 131] operating at cryogenic temperatures as intended, the shifts of the resonance wavelength and modifications of quality factors of individual elements have to be compensated. Therefore, design and material properties, in particular the refractive index, must be known a priori.

In this section, a method to measure the temperature-dependent refractive index in ultrasmall volumes is presented on the basis of Ref. [252].

12.2.1 *Experimental Method*

For measuring the refractive index shifts, at first the resonance shift of a photonic crystal cavity has to be measured. This resonance shift has two contributions: the change of the refractive index [287] and the change of the geometry due to thermal expansion [288]. In most materials thermal expansion can be neglected and the only relevant process is the change of the refractive index n [287, 288]. FDTD simulations unravel that the change of the resonance with the refractive index does not depend on the thickness of the photonic crystal slab (Fig. 11.14). Thus, a change in the resonance wavelength is a very precise measure for changes in the refractive index, even if the exact geometry is not known.

Because for the resonance only the material occupied by the cavity mode is relevant, this allows refractive index measurement in ultrasmall volumes. For example, when a 60 nm thick membrane with lattice constant of $a = 209$ nm is used, the L3 cavity mode concentrated on about seven unit cells corresponds to a material volume of 14 attoliter (see Fig. 11.12). Hence in contrast to other methods allowing refractive index measurements in bulk material [289] or thin membranes [290], this allows measurements in ultrasmall volumes on the order of 10 attoliters. This is probably the smallest volume in which refractive index measurements can be performed.

To control the cavity temperature within a wide range, the sample can be mounted on the cold finger of a continuous-flow helium cryostat. By changing the helium flow and additional heating this allows to precisely regulate the temperature in the range between 5 and 300 K. In parallel, the spectral position of the cavity resonance has to be measured. For this, using the intrinsic material fluorescence under incoherent excitation (Section 11.5.1) is unsuitable, as the excitation of fluorescence requires relatively strong lasers (in the order of 10 µW), which would unavoidably heat the PC cavity. To avoid this heating, the crossed polarization scheme from Section 11.5.2 is preferable. In this scheme only very low illumination power is needed ($P < 100$ nW). By doubling the power and comparing the position of the cavity resonance it can be guaranteed that the radiation has negligible influence on the cavity temperature on the order of $\Delta T < 1$ K.

12.2.2 Temperature Dependency of the Refractive Index of GaP

Figure 12.4 shows an example of the measured resonance wavelength of the fundamental mode of a photonic crystal cavity with hole radius $r = 63$ nm fabricated in a 59 nm thick GaP membrane, measured at various temperatures between 5 K and near room temperature. To verify the reproducibility, the data were taken in two subsequent cooling and heating cycles. From the change of the resonance wavelength the refractive index n can be calculated according to the relation

$$n(T) = n_{300K} + [\lambda(T) - \lambda_{300K}] \cdot a_{55}, \qquad (12.3)$$

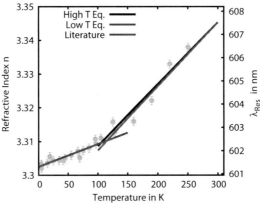

Figure 12.4 The measured cavity resonance wavelength λ_{Res} (right axis) and the calculated refractive index of GaP (left axis) between 5 and 300 K. To verify the reproducibility, the data were taken in two repeated cooling and heating cycles. The solid line corresponds to Eq. 12.4 for temperature above 100 K and Eq. 12.5 for temperatures below 100 K. The data above 100 K are in good agreement with the literature, e.g., Ref. [287]. From Ref. [252], © 2013 IOP.

with the refractive index at room temperature $n_{300K} = 3.34537$ from Table 11.1, the measured resonance wavelength at room temperature $\lambda_{300K} = (608.20 \pm 0.01)$ nm and the slope $a_{55} = 1/(163.36 \pm 0.01)$ nm gained from the FDTD simulations for a slab thickness t of 55 nm (Fig. 11.14). The resulting refractive index is shown in Fig. 12.4, together with fitted curves.

Above 100 K the refractive index changes linearly according to

$$n(T) = 3.290 \pm 0.001 + (180 \pm 20) \cdot 10^{-6} \cdot T/K. \quad (12.4)$$

This is in very good agreement with previous values for the refractive index of GaP at temperatures above 100 K from Ref. [287]. Interestingly, the slope decreases at a temperature of $T \approx 100$ K and the refractive index follows

$$n(T) = 3.302 \pm 0.001 + (67 \pm 7) \cdot 10^{-6} \cdot T/K. \quad (12.5)$$

Fitting the entire dataset shown in Fig. 12.4 with an exponential function is also possible, but the two-slopes model represented by Eqs. 12.4 and 12.5 is clearly more accurate. Although no low-temperature data for the optical refractive index were previously available in the literature, the kink at about 100 K is qualitatively in agreement with measurements of the low-frequency dielectric constant [239].

12.2.3 *Influence of the Temperature on the Quality Factor*

According to FDTD simulations the Q factor is almost unaffected by a small change of the refractive index. Surprisingly, one finds experimentally that the decrease of the temperature is followed by an increasing Q factor (Fig. 12.5). Starting with a value of $Q = 560 \pm 50$ at room temperature, the Q factor of the fundamental mode increases linearly with decreasing temperature. The highest measured Q factor of $Q = 1150 \pm 50$ is more than two times higher than the initial value at room temperature. This behavior can be attributed to reduced material absorption [291, 292] at low temperature due to freeze-out of phonons. This dependency highlights the importance of considering not only the desired operation wavelength, mode volume, and Q factor, but also the operation temperature of photonic crystal cavities.

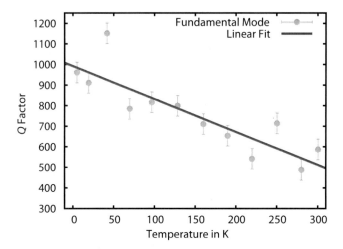

Figure 12.5 The measured Q factor of the fundamental cavity resonance increases with decreasing temperature. The highest measured value of $Q = 1150$ at $T = 42$ K is about two times higher than the value at room temperature. From Ref. [252], © 2013 IOP.

12.3 Thermo-Optical Switching

The shift of the PC cavity resonance as a function of temperature can also be used to implement integrated tunable filters and thermo-optical switching. The cavity is coupled to waveguides building an optical filter similar to Section 12.1. Close to resonance, already a small change of the refractive index leads to a large change in the transmission properties. This is almost ideal to build tunable filters or optical modulators and switches. Photonic crystal based optical switches have been demonstrated in the infrared [293–296]. There, a modification of the index of refraction via optical free carrier injection is preferable because of the faster switching speed needed in telecommunication technology. In what follows, switching in the visible spectral range is exploited. Since the size of integrated optical devices scales with the third power of the wavelength, cavity-based switching can be achieved with much smaller cavity volumes. As demonstrated in the following on the example of photonic crystal cavities in GaP, this in principle leads to fast switching speeds even via thermo-optic effects.

(a)   (b)

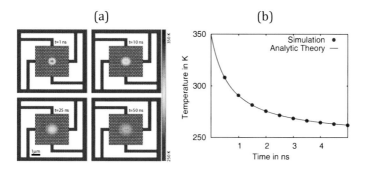

Figure 12.6 (a) Snapshots of the simulated temperature distribution 1 ns, 10 ns, 25 ns, and 50 ns after injection of a Gaussian shaped heat pulse. (b) Cavity temperature after a delta-shaped pulse. The figure shows the result of the simulations (dots) and the analytical solution (solid line) with effective thermal diffusivity. From Ref. [252], © 2013 IOP.

12.3.1 *Theoretical Predictions*

To estimate the energy required to switch the device, a specific heat of GaP of 410 J/(kg K) [297, 298] at operation temperature of 250 K is assumed. Assuming that the cavity volume is heated by a short laser pulse and using the experimental data from Section 12.2, one finds that the cavity resonance shift directly after the laser pulse is 1.3 pm per fJ of deposited energy. For cavities with moderate Q factors of $Q \sim 10\,000$ the temperature-induced resonance shift is on the order of the width of the resonance, when depositing 48 fJ. Thus, thermo-optical switching with an average energy consumption of less than 25 fJ per bit is feasible. To further reduce the energy required for switching, the operation temperature might be further reduced.

In the device thermal radiation can be neglected and the dynamic behavior of the temperature T is governed by the heat equation

$$\frac{\partial}{\partial t} T(\mathbf{x}, t) = \alpha(\mathbf{x}) \triangle T(\mathbf{x}, t), \qquad (12.6)$$

where the thermal diffusivity $\alpha(\mathbf{x})$ at position $\mathbf{x}$ equals the diffusivity in GaP $\alpha_{GaP} = k/\rho c_p = 66 \times 10^{-3}\ \mu m^2/ns$ [299] in the membrane, while it is zero in the holes. To investigate the switching dynamics, two-dimensional finite-difference simulations of the heat transport within the photonic crystal membrane and the mechanical support can be performed.

In detail the Crank–Nicolson method [300] can be used, and the remaining sparse system can be solved by the conjugate gradient method [301]. Here, Drichlet boundaries simulated the heat reservoir of the bulk material, whereas Neumann boundaries simulate the PC holes inhibiting the heat conduction process. Figure 12.6(a) shows the resulting temperature distribution in the PC slab at different times after the injection of a 3.7 pJ heat pulse. It is clearly visible that in the case of a single pulse the comparable small thermal conductivity of the membrane holders has no effect. To estimate the switching speed, the temperature evolution at the cavity position has to be evaluated [Fig. 12.6(b)]. This temperature evolution can perfectly be fitted with the analytical solution of the heat equation for a two-dimensional slab

$$T(t) = \frac{1}{4\pi \alpha_{eff}(t - t_0)}, \tag{12.7}$$

where the effective diffusion coefficient $\alpha_{eff} = (42.61 \pm 0.04) \times 10^{-3}\,\mu m^2/ns$. Remarkably the value of the coefficient cannot be explained by an average diffusivity deduced from the PC filling factor $\beta = 0.8$, which is $\beta \alpha_{GaP} = 53 \times 10^{-3}\,\mu m^2/ns$. Furthermore, according to Eq. (12.7) the cavity temperature difference is decreased to $1/e$ of the initial value after the time $\tau = 1.2$ ns. Thus, the device promises switching speeds on the order of 1 ns. In contrast for longer pulses, where the dynamics are limited by the heat transport through the support springs one finds $\tau \approx 680$ ns. To increase the switching speed the structure might be partially covered with a thin gold layer to increase the thermal conductivity and thereby allow faster switching. Additionally, the operation temperature might be further decreased, resulting in an improved ratio between specific heat and diffusivity.

12.3.2 Experimental Implementation

To experimentally verify the simulations, a cavity resonance can be probed with the crossed polarization method described in Section 11.5.2 to avoid the more complicated filter configuration with two waveguides. Here, a spectrometer can be used as a monochromator and light at the cavity wavelength can be detected with a fast avalanche photodiode. This allows for time resolution of 1 ns and

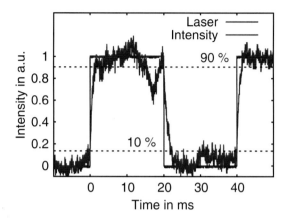

Figure 12.7 Real-time measured reflected intensity at the resonance wavelength when applying the heating laser for 20 ms intervals. The achieved contrast between on state and off state is above 13 dB, while the 10%–90% switch-on time is 1.2 ms. Switch-off is 2.2 ms. Negative intensities are due to correction for the detector background. From Ref. [252], © 2013 IOP.

below. In case of GaP, the slab material can be pumped with a switchable blue 405 nm laser. The light is efficiently absorbed and thus heats the cavity.

The observable contrast when switching the heating laser on and off exceeds 13 dB. Nevertheless, switching times on the order of about 1 ms (Fig. 12.7) are observable for laser powers of about 73 μW. This is over three orders of magnitude slower than predicted. Furthermore, the required heating power is several orders of magnitude above the predicted value. The large deviation originates in the parasitic heating of the entire substrates. A major fraction of the heating laser is absorbed in the silicon substrate, rather than in the membrane. Thus, the whole sample is heated up, which requires more energy and is much slower. Using transparent substrates or a different heating mechanism, these problems might be solved in future experiments. Nevertheless, the presented configuration is already suitable to perform active control of the membrane temperature [302].

PART IV

COUPLING OF QUANTUM SYSTEMS TO OPTICAL MICROSTRUCTURES

Introduction

In Part IV, the coupling of individual quantum systems to optical microstructures is studied. Exploiting the physics of (cavity) quantum electrodynamics, the hybrid systems allow otherwise unreachable, tailored light–matter interaction. For example, using the Purcell effect, enhanced single-photon emission from single quantum dots or NV centers is demonstrated and the strong coupling regime of cavity QED is reached with quantum dots.

More recently, on-chip quantum circuitry got into the focus of research [303]. In such quantum optical circuits several photonic components like waveguides and cavities are coupled to single emitters. Hereby, entire optical setups can be integrated on-chip [116]. In future experiments this approach might be further expanded to experiments with several optical active quantum systems, coupled on one single chip. Here, integrated hybrid systems improve existing interaction schemes, which have already been demonstrated and furthermore enable completely new cavity-mediated coupling schemes, as discussed in Chapter 15. Ultimately, these interaction schemes can be used to entangle several quantum emitters in a scalable and integrated fashion. This will certainly help explore the limits of quantum physics and potentially lead to applications in quantum information processing and metrology.

Following this outline, experimental realizations of the weak coupling regime are presented in Chapter 13. Here, quantum dots and NV centers are discussed, with particular attention on the required nanofabrication techniques. Thereafter, experiments reaching the strong coupling regime with quantum dots are reviewed in Chapter 14, before the prospects of on-chip integrated hybrids for entanglement generation are discussed in Chapter 15. This chapter represents also an outlook toward future experiments, as the proposed devices were not realized at the time of writing.

Chapter 13

Weak Coupling Regime

When an emitter is coupled to a low Q cavity, cavity damping dominates the dynamics and the system is in the weak coupling regime. As discussed in Section 4.4.5, in the weak coupling regime the spontaneous emission into the cavity mode is enhanced by the Purcell factor Eq. 4.120. To achieve Purcell factors $F \gg 1$, either the mode volume must be small or the quality factor must be high. These conditions are ideally fulfilled by the micro- and nanocavities introduced in the previous part.

13.1 Quantum Dots

By the year 2000 manufacturing of micro- and nanocavities became feasible and fluorescence from ensembles of quantum dots embedded into the cavity substrate was frequently used to characterize the devices [219]. It is not surprising that it took only a few more years until coupling of single Stranski–Krastanov-grown quantum dots to micropillar cavities [304, 305] and two-dimensional photonic crystal cavities [306, 307] was achieved. Here, the Purcell enhancement is used for efficient generation of single photons. One particular advantage of cavity enhancement is that it

Integrated Quantum Hybrid Systems
Janik Wolters
Copyright © 2015 Pan Stanford Publishing Pte. Ltd.
ISBN 978-981-4463-82-9 (Hardcover), 978-981-4463-83-6 (eBook)
www.panstanford.com

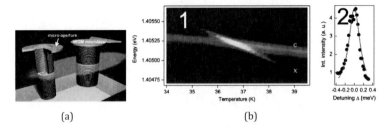

(a)   (b)

Figure 13.1 Integrated cavity QED experiment. (a) Schematic view of device design. An electrically pumped, in-plane emitting WGM microlaser (right) excites a high-Q vertically emitting QD micropillar (left). A micro-aperture suppresses stray light from the WGM laser when detecting the photoluminescence (PL) signal from the micropillar. (b1) Temperature-dependent PL map (linear color scale) of the micropillar cavity (diameter = 2.5 μm, Q = 10,500) under excitation from the electrically pumped WGM microlaser. Enhancement of emission from QD exciton X at resonance with the fundamental cavity mode C of the micropillar reveals light–matter interaction in the weak coupling regime. (b2) Integrated emission intensity of X as a function of detuning Δ. A fit (solid line) to the experimental data allows to determine a Purcell factor of $F = 4.1 \pm 0.6$. From Ref. [308], © 2013 Wiley-VCH Verlag GmbH & Co.

occurs only on a single, well-defined spatial mode, the cavity mode. Thus, natural coupling to single-mode waveguides and fibers is possible and today many experiments with single quantum dots like the ones shown in Fig. 13.1 make use of nanostructures to enhance the collection efficiency.

A second property of the Purcell enhancement is its spectral selectivity. Only decay channels with a spectral overlap with the cavity modes are enhanced, corresponding to a relative suppression of phonon side-bands. This is a particular issue when spectrally equal or similar photons are required for multi-photon interference experiments [152, 309–312] or interfacing of QDs with other quantum systems [74, 313–315], which might be atoms, photons from entangled photon pair sources, or other quantum dots.

The spatial selectivity of the Purcell effect, manifested in the form factor $\mathcal{F}(\mathbf{x})$ (Eq. 4.103), becomes critical during the interfacing of several quantum dots embedded in an on-chip photonic circuit. Here, several quantum dots with well-defined properties must be located at specific cavity modes. To achieve this, the random positioning of the Stranski–Krastanov-grown quantum dots is a major challenge. The probability of finding the number of *N* quantum dots with high spectral quality scales exponentially in *N*, making it unlikely to find a specific configuration by chance. Thus, much effort is spent to fabricate cavities on demand at the position of pre-characterized quantum dots or to use site-controlled growth methods [316]. Nevertheless, side-controlled quantum dots which might be an ultimate solution to the positioning problem do not achieve the high spectral stability of Stranski–Krastanov-grown ones and thus are not yet suitable for advanced experiments on the interaction of several QDs.

13.2 Color Centers in Diamond

To optically couple single-color centers to cavity modes, in general two approaches are possible. In the top-down approach, optical microstructures are fabricated from substrates already containing color centers, very similar to the experiments with single quantum dots described in the previous section. Alternatively, in the bottom-up approach, diamond nanocrystals containing single-color centers are placed inside the cavity mode in a postprocessing step. The first approach is easily applied on a large scale, while the placement of the individual emitters is difficult to control and is random, as discussed above. In contrast, the latter approach has the advantage of being intrinsically deterministic, while it is not suitable for fabrication on large, industrial scale. Both approaches are presented in the following sections 13.2.1 and 13.2.2.

13.2.1 *Top-Down Integration*

For the *top-down* integration of single-color centers into photonic micro- and nanostructures, two routes can be followed. The entire device can be made out of either diamond or a matrix material in which nanodiamonds with color centers are embedded.

Diamond Strucutres Fabrication of nanostructures out of mono-crystalline diamond is very difficult to achieve because of the hardness of the material. In recent years several research groups spent much effort in the development of suitable structuring processes and diamond microring resonators [117, 317], photonic crystals [318–320], and waveguides [321] could be realized. Despite the tremendous progress, in general the material quality degrades during processing and single defect centers could only be observed in few experiments [317, 319]. A Purcell factor of $F = 70$ was achieved by a single NV center coupled to an L3-type photonic crystal cavity with $Q \sim 3000$ and $V \sim (\lambda/n)^3$ [319]. To achieve deterministic coupling, it is possible to implant single nitrogen atoms and thereby generate NV centers at well-defined positions [322]. Today, the positioning precision in the implantation process is not yet high enough to implant ions into the micro-size cavity, but this may change in the future with progress on ion trapping [323, 324] or nano-apertures [325].

Polymer Structures with Nanodiamonds The alternative route is the fabrication of nanostructures from a nanodiamond-containing matrix, e.g., a polymer. As discussed in Section 10.3.1, the use of acry-late polymers and state-of-the-art direct laser writing techniques makes fabrication of such structures comparably easy. Here, it is important that the color centers, e.g., NV centers, are extremely photostable and do not degrade when exposed in the direct laser writing process. Figure 13.2 shows a disk resonator with a diameter of 20 µm and thickness of ∼1.2 µm written in a nanodiamond-containing polymer. By performing a confocal scan, individual NV centers sitting on the outer rim where the mode is expected to have field maxima can be identified. The measured spectrum clearly shows a multi-peak structure. Each peak corresponds to a cavity resonance, where the spontaneous emission is Purcell enhanced. To prove the single emitter character the autocorrelation function $g^{(2)}(\tau)$ is measured as illustrated in Fig. 13.2(d).

A more sophisticated experiment is shown in Fig. 13.3. A nanodiamond-containing resonator is coupled on-chip to a dielec-tric waveguide with a width of 1.8 µm and a length of about 40 µm. Here the three-dimensionality of the DLW process is exploited to demonstrate not only emitter–resonator coupling but also on-chip circuitry.

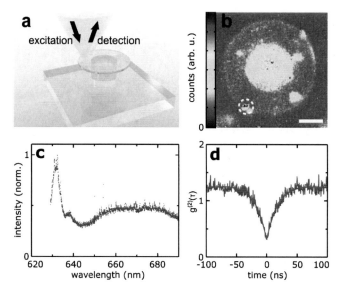

Figure 13.2 Single NV centers in a DLW disk resonator. (a) Detection scheme. The NV-containing resonator is scanned through a confocal detection volume. (b) Scanning confocal image of the resonator. Scale bar is 5 μm. Bright spots can be identified as single NV centers. (c) Spectrum taken of the center encircled in (b). Ripples on the phonon side-band from 640 to 700 nm correspond to individual cavity modes. The peak at 630 nm stems from the fluorescence background from the polymer. (d) Autocorrelation function measured on the same NV center. The dip at $\tau = 0$ clearly proves the single emitter character. From Ref. [198], © 2013 Nature Publishing Group.

Again, when individual NV centers in the outer rim of the resonator are excited by a focused laser, single-photon emission is resonantly enhanced into the cavity modes. Instead of directly detecting the photons, light is coupled into the waveguide being in the near field of the cavity. As the coupling occurs in the middle of the waveguide, the photons can either travel to the left or right output port, where they are collected and finally detected. In this configuration the waveguide acts as the beam splitter in an integrated Hanbury Brown and Twiss setup. The background-corrected autocorrelation function between photons detected in the left and right port indeed shows the expected anti-bunching behavior.

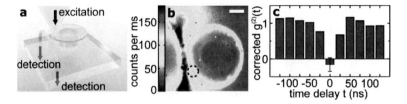

Figure 13.3 On-Chip Quantum Circuitry with DLW structures. (a) Illustration of the chip and the experiment. Individual NV centers in the outer rim of the resonator are excited by a focused laser beam and single-photon emission is routed through the waveguide toward the detectors. (b) Map of the intensity measured at the waveguide output ports when scanning the laser over the structure. The waveguide and the resonator are clearly visible by enhanced fluorescence. Individual bright spots correspond to NV centers. Scale bar is 5 μm. (c) Autocorrelation function between the waveguide outputs, when the spot encircled in (b) is excited. The anti-bunching below zero at $\tau = 0$ is an artifact of the applied background correction. From Ref. [198], © 2013 Nature Publishing Group.

In the future, these chip-integrated quantum devices might be used for even more complex integrated experiments. Beam splitters, interferometers, and other components, like three-dimensional multi-port beam splitters [326] are immediately feasible. However, this scalability does not apply to the randomly placed emitters. Here, on-demand fabrication on samples with known emitter positions might partially solve this problem. Another method might be to use structured templates, where nanodiamonds with NV centers are placed in arrays. Although these techniques in principle might allow upscaling to several emitters, the spectral instability of nanodiamonds is still an open task.

13.2.2 *Bottom-Up Integration*

In the bottom-up approach, pre-selected diamond nanocrystals containing single defect centers are coupled to prefabricated photonic nanostructures. The required nano-assembly is done with modern nanopositioning techniques in scanning electron microscopy (SEM) [327, 328] or atomic force microscopy (AFM) [271, 329, 330], allowing to position the diamond nanocrystals with high precision, down to a few nanometers. Following this approach, several attempts have been pursued to couple NV centers to plasmonic or dielectric resonators, like dielectric mircrospheres [112, 331], microtoroids [332], photonic crystal cavities [128, 130, 132], and plasmonic nano-antennas [330, 333] and thereby enhance the spontaneous emission rate.

The next paragraph gives an overview on the AFM pick-and-place technique based on Ref. [330], which is followed by experimental results on coupling of NVs to photonic crystal cavities based on Refs. [128, 129].

AFM Pick-and-Place Technique The *pick-and-place* technology allows to pick up individual diamond nanocrystals with single NV centers from a cover slip and place them on almost arbitrary structures by means of an atomic force microscope. A sketch of the experimental setup is shown in Fig. 13.4(a), consisting of a confocal microscope with an AFM on top.

Nanodiamonds are prepared in a usual procedure: a commercial suspension of 25 nm-sized diamonds is spin-coated onto a glass cover slip. An estimate 5% of these nanodiamonds host an NV center. Then, the sample is optically excited through the microscope and fluorescence light is collected with the same objective and detected by two avalanche photodiodes in an HBT configuration. The

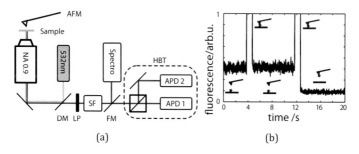

(a)   (b)

Figure 13.4 The pick-and-place technique. (a) Sketch of the used setup consisting of a confocal microscope with an atomic force microscope (AFM) atop (see Fig. 7.4 for details on the optical setup). (b) Fluorescence signal during pick-up. When the cantilever is pressed onto the sample fluorescence increases because of optical excitation of the cantilever. In case of successful pick-up it drops to background level when the cantilever is retracted. Adapted from Ref. [330], © 2013 AIP.

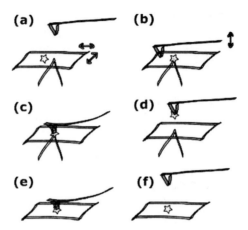

Figure 13.5 Cartoon illustrating the pick-and-place procedure. (a) A nanodiamond with desired properties is found by scanning the sample. (b) By scanning the AFM tip the diamond in the focus is identified. (c) The AFM tip is pressed onto the diamond. (d) The diamond sticks to the AFM tip and is transferred to the target sample. (e) The cantilever with attached diamond is pressed onto the sample to release it. (f) The properly positioned diamond is lying on the target structure. From Ref. [330], © 2013 AIP.

photon detection times and the trigger signal of the excitation laser are recorded with a photon counting system in time-tagged time-resolved (TTTR) mode, in which all detection events are stored with high timing accuracy. From this data, the correlation function $g^{(2)}(\tau)$ as well as lifetime transients can be obtained. Furthermore, spectra of the NV centers can be measured [Fig. 13.5(a)]. After identifying a proper nanodiamond, the metal-coated tip of the AFM is used to pick it up by pressing the cantilever onto it [Fig. 13.5(b-c)]. To control this pick-up process, the fluorescence intensity is monitored simultaneously. While the cantilever is approached, the intensity increases drastically because of additional fluorescence from the metal-coated cantilever. When it is retracted again, the fluorescence drops to the previous level in case of an unsuccessful attempt, or to background level in case of successful pick-up [Fig. 13.4(d)]. When the diamond sticks to the AFM tip it can be transferred to another substrate containing the target structure.

For releasing the diamond, a similar procedure is followed. The cantilever with the diamond on the tip is pressed onto the target structure [Fig. 13.5(e-f)]. In case of successful placement, a subsequent scan in tapping mode shows a new object, the nanodiamond. Here it is helpful to carefully scan the target area in tapping mode prior to the placement, to easily identify the nanodiamond afterwards. Nanomanipulation with the AFM tip then allows fine-positioning of the nanocrystal and launching of single excitations at arbitrary positions near photonic components.

While the pick-up process works with unity efficiency, the placing efficiency is estimated to be on the order of 10%. The complex interplay between surface forces allowing the pick-and-place technique is not completely understood, and thus precise rules for improving the success probability cannot be given. However, two general rules can be given. On the one hand, the radius of curvature of the AFM tip should not be too small to have a high probability for the diamond to stick on the apex rather than the side (Fig. 13.6). On the other hand, the tip must be small enough to identify individual nanodiamonds. Second, brittle materials are unfavorable as they break when pressing the cantilever onto the sample. In experiments, Au and Pt/Ti-coated cantilevers have proven to fulfill these requirements.

With the pick-and-place technique, individual nanodiamonds containing exactly one single NV defect center can be precisely positioned at a desired position on a target sample. This can be

Figure 13.6 Scanning electron microscope image of an AFM tip used for pick-and-place. The apex is flattened by pressing it to the surface during pick-up process. Scalebar is 1 μm. From Ref. [330], © 2013 AIP.

a plasmonic structure [129, 334], optical fiber [335], or photonic crystal cavity as described below. By using this technique, one can guarantee that there is exactly one diamond on the desired position or even on the whole sample.

Integration of NV Centers into Photonic Crystal Cavities As discussed earlier, the optical properties of the NV centers are not ideal: The coupling strength to the electromagnetic field is small compared with other systems like quantum dots [336]. Another problem is the small Debye–Waller factor. Even at cryogenic temperatures only a small fraction of about 4% of the radiation is emitted into the Fourier-limited zero-phonon line (ZPL) [337].

These problems can be overcome by coupling the NV centers to optical microcavities [337, 338], where the spontaneous emission (SE) rate is enhanced by the Purcell effect. Increasing the SE rate into the ZPL will allow the generation of a large number of indistinguishable single photons [338] needed, e.g., for linear optics quantum computation [170].

Several early attempts have been pursued to couple NV centers to cavities such as microsphere resonators [112, 331], micro-toroids [332, 339], or photonic crystal cavities [132, 271]. Among these, photonic crystal cavities have the earlier discussed advantage that they can be fabricated with high quality factors as well as small mode volumes, both being figures of merit for obtaining a large Purcell enhancement. Furthermore, photonic crystal cavities can be easily integrated into more complex systems of coupled cavities and waveguides. Thus, the experimental realization of the coupling of the ZPL of an NV center in a nanodiamond to a single mode of a photonic crystal cavity is one of the most crucial steps for integrated diamond-based quantum technology in a bottom-up approach.

The early attempts to achieve this step were based on silicon nitride cavities. There, material background fluorescence dominated, and single-photon emission from single NVs could not be observed [267, 271]. In 2010 several research groups independently solved this problem by using GaP as cavity material [128, 130, 132]. The following paragraphs are based on results from Ref. [128].

At first, a suitable diamond containing a single NV center is selected and pre-characterized on a cover slip. This is done using

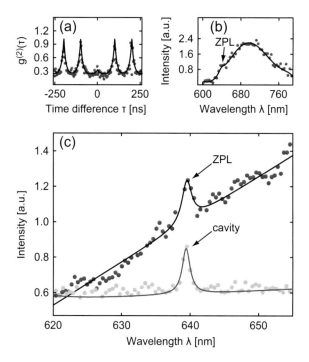

Figure 13.7 Pre-characterization of the Constituents. (a) Pulsed autocorrelation measurement from a nanodiamond on a glass cover slip, clearly revealing the single photon character of the fluorescence. The black curve is a fit to the data. (b) Full spectrum of the emission from a single NV center in a nanodiamond on the GaP substrate under cw excitation. The solid line is a guide to the eyes. (c) Upper curve: Zoom in on the ZPL of the same NV center as in (b). The solid line is a fit to the data. Lower curve: Corresponding measurement of the cavity fluorescence spectrum. The solid line is a fit to the data, showing almost perfect matching with the spectral position of the ZPL. For better visibility an offset of 0.5 is added to the data. From Ref. [128], © 2010 AIP.

a confocal microscope (NA 1.35) with a HBT setup under pulsed excitation with a 532 nm laser at a repetition rate of 10 MHz. Figure 13.7(a) shows the measured $g^{(2)}$-function of a single nanodiamond. The peaked structure represents the repetition time of the pulsed laser. In contrast to what is expected from classical light there is no peak at $\tau = 0$ ns, showing that the diamond contains only a single NV center. Using the above described pick-and-place technique this preselected diamond is picked up with an atomic force microscope and preliminarily placed on the GaP surface, close to the photonic crystal cavities. Here the diamond is further characterized. In particular, a spectrum is measured [Fig. 13.7(b)] and the position of the ZPL at 639.5 nm is determined [Fig. 13.7(c)].

Then a cavity with a resonance wavelength of 642.9 nm and a Q factor of $Q = 1003$ is selected and actively tuned by repeated local laser heating of the cavity to the ZPL of this diamond as described in Section 11.6. Figure 13.7(c) shows the fluorescence spectra of the bare cavity after tuning and the fluorescence from the NV center. Note that the cavity resonance wavelength matches the ZPL almost perfectly.

In a last step, the very same nanodiamond is picked up again using the AFM and is placed in the center of the cavity now [Fig. 13.8(a,c)]. Since the field strength of the cavity mode decays exponentially outside the slab, it is crucial to place the NV center as close to the surface as possible [132]. In the AFM images the diamond shows a height of about 35 nm, meaning that the NV center is less than 35 nm above the surface. Figure 13.9(a) shows the measured spectra of the cavity with and without the nanodiamond, both taken at 300 μW excitation power. The pronounced enhancement of the ZPL emission at 639.5 nm is striking. The autocorrelation measurement [inset in Fig. 13.9(a)] proves that the light emitted from the cavity has a strong nonclassical behavior, revealing that the fluorescence mainly originates from the NV center. The peaks around 610 nm stem from higher-order modes with smaller Q factors. As no influence of the NV center on the fluorescence intensity is expected below 620 nm, the two spectra were normalized with respect to the emission at 600 nm to compensate for a slightly different alignment. This allows to reveal which part of the hybrid system's fluorescence has its origin in the NV center by subtracting the normalized bare cavity background.

To quantify the observed enhancement in the ZPL, the spectrally resolved spontaneous emission enhancement at the cavity wavelength F^* is calculated by comparing the ZPL emission in the cavity and on the bare substrate. In case of vanishing phonon side bands, this quantity F^* would equal the overall Purcell factor F. When calculating F^* care has to be taken, as any change in the dielectric

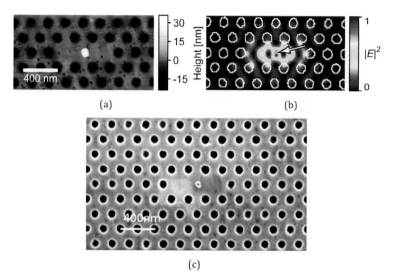

(a)   (b)

(c)

Figure 13.8 A single NV center in a nanodiamond coupled to a photonic crystal cavity. (a) AFM image of the L3 GaP cavity with a nanodiamond of height 35 nm located close to the center. The lattice constant is 200 nm. (b) Simulated (FDTD) electric field profile of the fundamental mode of the cavity. The geometry was adapted from the AFM image in (a). The arrow indicates the position of the diamond. (c) The corresponding scanning electron microscope image of the same region. (a) and (b) from Ref. [128], © 2010 AIP, (c) from Ref. [129], © 2012 WILEY-VCH Verlag GmbH & Co.

environment also influences the corresponding emission properties such as the overall spontaneous emission rate [132]. However, the spectral shape changes only near the cavity resonances. This allows to normalize the measured spectra to a broad spectral region above and below the ZPL. Afterwards, a Lorentzian is fitted to the data, resulting in normalized peak intensities of $I^*_{sub} = 1.2$ and $I^*_{cav} = 14.9$ on the substrate and on the cavity, respectively [Fig. 13.9(b)]. This yields the experimental Purcell enhancement $F^* = I^*_{cav}/I^*_{sub} = 12$. The difference to the maximal theoretical value $F = 61$, calculated from Eq. 4.120, is caused by non-ideal alignment. Neither the NV center is placed in the field maximum of the cavity mode, nor the dipole moment's orientation is optimized. This is a general problem of the bottom-up approach and is comparable to other results. Ref. [130] reports an enhancement by a factor of four, while

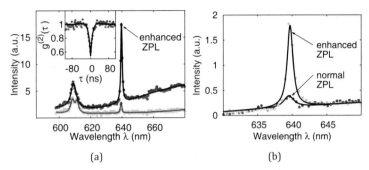

Figure 13.9 (a) Lower curve: Normalized spectrum of the cavity fluorescence without nanodiamond. The peaks around 610 nm stem from higher-order modes. The solid line is a fit to the data. Upper curve: Normalized fluorescence spectrum with nanodiamond. For better visibility an offset of 1 is added. The solid line is a fit to the data. Inset: Autocorrelation measurement of the cavity with nanodiamond. The dip at $\tau = 0$ ns proves the nonclassical behavior of the light emitted from the assembled system. (b) Comparison between the background corrected normal ZPL measured outside the resonator and the enhanced ZPL measured inside the resonator. (a) Adapted from Ref. [128], © 2010 AIP, (b) adapted from Ref. [129], © 2012 Wiley-VCH Verlag GmbH & Co.

in Ref. [132] gives $F^* = 5$. Only recently, Ref. [319] reported an achieved Purcell factor of $F^* = 70$ by using a top-down approach.

A further enhancement of the emission into the ZPL can be achieved by improving the Q factor or by performing experiments at cryogenic temperatures [319]. At 4 K about 4% of the light is emitted into the ZPL. In this case coupling to a similar photonic crystal cavity with a Q factor of 600 should allow channeling of more than 30% of the emission into the cavity mode.

The bottom-up approach allows to demonstrate the deterministic coupling of the zero-phonon line of a single NV center in a nanodiamond to a photonic crystal cavity. This is a major step toward the realization of integrated quantum optical devices. With the presented pick-and-place technique and the selective cavity tuning, even more complex systems, involving several cavities and emitters [340], can be assembled in a controlled way. Simple quantum gates, integrated on a single photonic crystal chip, are within reach when spectral diffusion on nanodiamonds can be controlled.

13.3 Applications of NV Centers in the Weak Coupling Regime

The weak coupling regime has several interesting applications. The most obvious and prominent one is the efficient single-photon generation. Figure 13.10 shows a sketch of a future ultrabright resonantly enhanced single-photon source. The proposed source consists of a nitrogen-vacancy center in nanodiamond and a photonic crystal cavity which is coupled via waveguide to a free-space grating coupler [383]. Such bright and spectrally narrow single-photon devices may not only demonstrate single-photon generation and routing on a photonic crystal chip [341], but are also key ingredients for many quantum optics experiments and technologies, like quantum cryptography. The enhanced photon emission of these hybrid devices can furthermore be used to improve the spin detection efficiency, potentially allowing single-shot readout.

In particular, cryogenic experiments may allow not only the standard ODMR technique discussed in Section 7.5.3, but also more sophisticated schemes as demonstrated on quantum dots [303, 342]. For example, Ref. [343] suggests that resonant scattering by the NV center might lead to an impedance mismatch between cavity and waveguide. This mismatch can be detected by measuring the

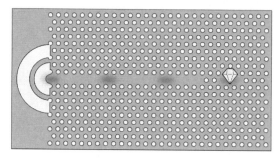

Figure 13.10 Sketch of a future cavity-enhanced single-photon source. An NV-containing nanodiamond is placed in a photonic crystal cavity to resonantly enhance the single-photon emission (right-hand side). Subsequently, the red single photons from the resonantly enhanced ZPL at 637 nm are coupled to a photonic crystal waveguide and scattered out of the photonic crystal plane by a grating coupler (left hand side).

reflectivity of the system at a specific spin-dependent NV transition. Using this scheme, a readout fidelity of 99% is predicted for a uncoupled cavity Q factor of $Q_{un} = 3000$ and an overall Purcell factor of $F^* = 3$.

Although this scheme appears promising, its application might be inhibited by spectral diffusion, in particular in nanodiamonds, making the observation of stable spin-dependent absorption lines impossible.

Chapter 14

Strong Coupling

As discussed in Section 4.4 the strong coupling regime of cavity QED is reached when the emitter–cavity coupling rate g_{cav} dominates over the cavity loss rate κ and other decay modes of the emitter γ, i.e., $g_{cav} > \kappa/2, \gamma$. The strong coupling regime is not only ideal to test the field quantization formalism and the resulting Jaynes–Cummings model [56, 344], but also the basis for many scalable quantum computation and networking protocols [345–347]. Here, the cavity mediates deterministic light–matter interaction on the single-photon single-emitter level, allowing the realization of quantum gates and interfaces between stationary and flying qubits. In classical information processing, strong coupling can be used to achieve ultrafast all-optical switches with ultralow switching energies, reaching the one-photon level [348]. Thus, strongly coupled solid state emitters promise to provide a scalable quantum and classical optical information processing platform.

The strong coupling regime was reached with atoms coupled to superconducting and optical cavities first [57], and later on with quantum dots in photonic crystal cavities [336, 349–351]. In the following section, recent strong coupling experiments with quantum dots in photonic crystal cavities are reviewed, while prospects of strong coupling with defects in diamond are discussed later on.

Integrated Quantum Hybrid Systems
Janik Wolters
Copyright © 2015 Pan Stanford Publishing Pte. Ltd.
ISBN 978-981-4463-82-9 (Hardcover), 978-981-4463-83-6 (eBook)
www.panstanford.com

14.1 Strong Coupling of Quantum Dots to Microcavities

Quantum dots provide several important prerequisites for achieving the strong coupling regime. First, the short radiative exciton lifetime of about 1 ns goes hand in hand with a large dipole moment (Table 5.2), allowing large cavity couplings. Second, narrow-band optical emission from the exciton is the dominating decay channel and hence the decay rate into non-cavity modes γ is small and can be neglected. Third, as discussed above, integration into low-volume high-Q (low κ) cavities is possible with today's technology. Indeed, several groups achieved strong coupling of single quantum dots by the year 2004, using photonic crystal cavities [349], microdisks [350], and micropillar cavities [351].

Figure 14.1 shows results from the latter experiment. A layer of natural $In_{0.3}Ga_{0.7}As$ quantum dots [352] is embedded in a low mode volume micropillar cavity. These quantum dots have a comparably large size of about 100 nm and therefore have a large dipole moment. The pillar cavity used for the strong coupling experiment provides lateral index confinement within its 1.5 μm diameter and bandgap confinement by GaAs/AlAs DBRs in the vertical direction.

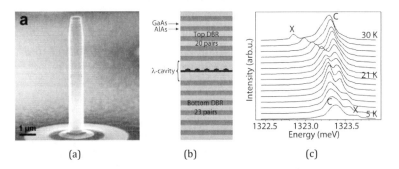

(a)   (b)   (c)

Figure 14.1 Strong Coupling Experiment with Quantum Dots. (a) Scanning electron micrograph of a micropillar cavity with 0.8 μm diameter. (b) Sketch of the micropillar cavity of size λ/n. Lateral confinement is achieved by the index contrast to the surrounding vacuum, while vertical confinement is achieved by AlAs/GaAs DBRs. (c) Temperature-dependent luminescence spectra of QD exciton (X) and cavity mode (C). The exciton is tuned through the cavity resonance. At about 21 K the anti-crossing is visible, proving the strong coupling. Adapted from Ref. [351], © 2004 Nature Publishing Group.

Here, the lower DBR has slightly more layers than its counterpart on the top of the device to achieve preferential emission into the latter direction (cf. Fig. 14.1(b)).

With this structure it is possible to experimentally achieve Q factors of $Q \sim 9000$, corresponding to $\hbar\kappa \sim 25$ µeV, while the mode volume is calculated to be on the order of $V \sim 15(\lambda/n)^3$. By off-resonant excitation into the conduction band it is possible to generate fluorescence in the quantum dots. At a temperature of 5 K all QDs in the cavity have slightly different emission wavelength and it is possible to spectrally select emission of a single quantum dot matching almost the cavity resonance. Now, one can use the fact that the cavity resonance is almost unchanged when heating up to about 30 K, while the exciton line shifts by many linewidths. This allows to study the system when QD and cavity come into resonance [Fig. 14.1(c)].

When QD and cavity are strongly coupled an avoided crossing is observed. This anti-crossing is even better visible when looking at the fitted peak positions shown in Fig. 14.2(a). From the Rabi splitting at zero detuning the coupling constant can be deduced to be $\hbar g_0 = 80$ µeV. This is about a factor of 3 above the strong coupling threshold, i.e., $g_0 \approx 3\kappa$. In contrast, in the weakly coupled

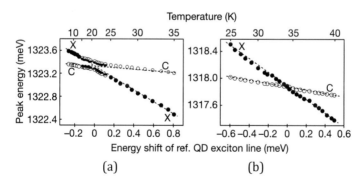

Figure 14.2 Photoluminescence peaks of the QD and cavity mode when varying the temperature. The lower axis gives the corresponding shift of a reference QD not coupled to the cavity. (a) In the strongly coupled system the avoided crossing between exciton (X) and cavity mode (C) is clearly visible. (b) In the weakly coupled reference system both lines cross. Adapted from Ref. [351], © 2004 Nature Publishing Group.

reference system, exciton and cavity can have the same frequency [Fig. 14.2(b)].

More recent experiments aim toward utilizing the strong coupling for switching [302, 348] or generation of nonclassical light [353, 354]. Future experiments might make use of the electron spin in charged quantum dots, where strong optical coupling provides an efficient spin–photon interface [355].

14.2 Strong Coupling with NVs in Diamond

As discussed in the previous section, several experiments have realized strong coupling with QDs. For color centers in diamond this was not possible so far. Here, two major hurdles are fabrication issues (cf. Section 13.2) and spectral diffusion (cf. Section 7.4). In the future these might be overcome, paving the way to strong coupling of defect centers.

In particular, for the NV centers, strong coupling is desirable, as it provides an excellent quantum memory with its long living coherent spin. Here, several schemes have been proposed to build a scalable quantum information processing platform with NV centers strongly coupled to cavities [46, 356–359]. In the following, the required cavity parameters to reach the strong coupling regime are estimated.

When neglecting nonradiative decay channels, the spontaneous emission rate into the NVs' zero-phonon line γ_{ZPL} can be calculated from the excited state lifetime $\tau = 1/\gamma_{tot} \sim 12$ ns in bulk diamond to be approximately $\gamma_{ZPL} = D/\tau = 4$ MHz, with $D = 0.05$ being the Debye–Wallert factor. From this the dipole moment can be estimated by using Eq. 4.23 and Eq. 8.21:

$$|\langle \hat{\mathbf{D}} \rangle| = 0.44 \, e a_0, \qquad (14.1)$$

where e is the elementary charge and a_0 is the Bohr radius. One can estimate that in case of perfect overlap of the dipole with the cavity mode the strong coupling condition $g > \kappa, \gamma$ is reached, when $Q > 3\sqrt{V/\left(\frac{\lambda}{n}\right)^3} \cdot 10^5$ and $V/\left(\frac{\lambda}{n}\right)^3 < 360$. This is illustrated in Fig. 14.3. Even for very small mode volumes of $V = 0.1 \left(\frac{\lambda}{n}\right)^3$ the quality factor must exceed $Q = 10^5$ to achieve strong coupling [360].

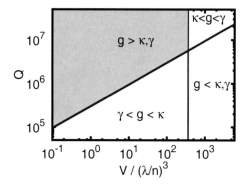

Figure 14.3 Cavity QED coupling regimes for NV centers. The vertical line indicates when $g = \gamma$, while the diagonal indicates $g = \kappa$. Strong coupling is reached only when $g > \kappa, \gamma$, corresponding to the shaded area. The calculations assume perfect overlap of the dipole with the cavity mode.

As the required high Q factors have not been demonstrated so far for visible light, it is questionable whether strong coupling will be achieved with NV centers in the near future. Probably more feasible is the strong coupling of other defects like SiV centers. These provide larger dipole moments, reducing the required cavity Q factor. Anyhow, also the weak coupling regime can be used to realize cavity-enhanced entanglement schemes as discussed in Chapter 15.

Chapter 15

Cavity-Enhanced Entanglement

Entanglement, or the existence of nonclassical correlations between distinct, potentially distant quantum objects, is one of the most intriguing aspects of quantum mechanics. The first discussed, and probably most famous, phenomenon arising from entanglement is the *EPR paradox*, published in 1935 by Albert Einstein et al. [361]. While the early discussions of this paradox remained on the philosophical level of Gedankenexperiments, it took about 30 years until it was put into a verifiable mathematical form in 1964, today known as the *Bell inequality* [362]. This and related formulations [363] have been experimentally tested many times since the early 1980s [364, 365]. None of these tests gave clear evidence for the incorrectness or incompleteness of quantum mechanics.

Apart from philosophical implications, today entanglement is considered a resource for quantum communication [74], computation [75], simulation [366], and metrology [367]. Thus, the generation of entanglement has been a holy grail of experimental quantum physics in the recent years. Although many experiments strikingly demonstrated entanglement on plenty of similar [122, 123, 364, 368–371] and dissimilar systems [121, 372–374], scaling to a quantum information processing network with many nodes, each

Integrated Quantum Hybrid Systems
Janik Wolters
Copyright © 2015 Pan Stanford Publishing Pte. Ltd.
ISBN 978-981-4463-82-9 (Hardcover), 978-981-4463-83-6 (eBook)
www.panstanford.com

having several quantum registers [375], could not be achieved yet. This scaling problem might be addressed by integrated solid state quantum hybrid systems based on quantum dots, superconducting circuits, or color centers.

Among the latter, the negatively charged nitrogen-vacancy (NV) center in diamond is regarded as one of the most promising candidates [28, 115, 118]. Technological progress of the recent years made it possible to integrate single NV centers into photonic crystal cavities (cf. Section 13.2). Furthermore, using short-range spin–spin interactions, entanglement between an NV's electron spin and adjacent nuclear spins [373] and between two NV centers [122] separated by 25 nm could be demonstrated. Recently, a probabilistic entanglement scheme [376] could be demonstrated for NV centers being 3 m apart [123].

The short-range interaction is well suited for quantum registers in a future quantum information processing node, while the probabilistic scheme might be applied to connect different nodes of future quantum information processing networks. Nevertheless, an additional medium-range interaction on the order of a wavelength is required for fast operations between individual registers of a quantum node. In this range an integrated optical platform for fast probabilistic, or deterministic entanglement [377] is promising.

Section 15.1 is dedicated to the recently demonstrated probabilistic measurement based entanglement scheme and possible improvements, while afterwards in Section 15.2 the fast deterministic scheme from Ref. [377] is discussed.

15.1 Probabilistic Entanglement

In probabilistic entanglement schemes, entanglement is generated not by an interaction between separate quantum systems, but by projective measurements. The advantage is that the individual quantum systems can be well isolated and thereby are much easier to control than interacting ones. This comes at the cost of an intrinsically probabilistic entanglement operation. Hence it cannot be guaranteed that entanglement is present on-demand. Anyhow, this issue can be overcome with high probability when the entanglement operation includes some heralding mechanism and the quantum systems provide sufficiently long coherence, i.e., entanglement storage times. In this case, the entanglement operation can be repeated until success and the entangled state is thereafter stored until needed. Thus, probabilistic schemes are well suited for most applications in quantum communication and quantum information processing.

In the recent years, several probabilistic entanglement schemes have been developed. Here the entanglement scheme presented in Ref. [376] is discussed. This scheme has already been demonstrated on NV centers [123] and is also suitable for integration into on-chip circuits.

15.1.1 A Heralded High-Fidelity Entanglement Scheme

By 2005 Barrett and Kok proposed a scheme for efficient high-fidelity entanglement with matter qubits and linear optics [376]. This scheme is based on identical L-type quantum systems, a beam splitter, and photon detectors, as illustrated in Fig. 15.1. The L-type

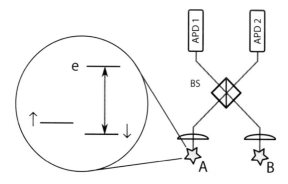

Figure 15.1 Illustration of the probabilistic entanglement scheme based on two individual equal L-type systems A and B with ground states $|\uparrow\rangle$ and $|\downarrow\rangle$. Each of the systems can be excited from state $|\downarrow\rangle$ to the excited state $|e\rangle$ by an optical π-pulse, from where they will emit single photons. The photons are collected (e.g., with high-NA lenses) and routed to a 50/50 beam splitter (BS). There, the information about which system emitted the photons is erased and the emission of a photon is subsequently detected by one of two photodiodes (APD1 and APD2).

quantum systems A and B provide two ground states $|\uparrow\rangle, |\downarrow\rangle$ and an excited state $|e\rangle$. The latter one is linked by an optical transition to one and only one ground state, e.g., $|\downarrow\rangle$, as opposed to Λ-type systems. It is assumed that both quantum systems can be initialized, manipulated, and read out individually.

To start the entanglement operation, both systems are independently initialized into the superposition $|\Psi_{0A/B}\rangle = 1/\sqrt{2}\,(|\uparrow\rangle + |\downarrow\rangle)$. This corresponds to the two-particle wave function

$$|\Psi_0\rangle = 1/2\,(|\uparrow_A\uparrow_B\rangle + |\uparrow_A\downarrow_B\rangle + |\downarrow_A\uparrow_B\rangle + |\downarrow_A\downarrow_B\rangle). \quad (15.1)$$

Now, the trick is to perform projective measurements, ruling out the states $|\uparrow_A\uparrow_B\rangle$ and $|\downarrow_A\downarrow_B\rangle$ from the wave function. This is done in the following way: An optical π-pulse is applied on the $|\downarrow\rangle \leftrightarrow |e\rangle$ transition, potentially bringing both systems into the excited state, from where a single photon is emitted. If at least one photon is detected, the state $|\uparrow_A\uparrow_B\rangle$ can be ruled out from the wave function.

To rule out also the other state $|\downarrow_A\downarrow_B\rangle$ first a ground state π-pulse is applied to flip the spin of each individual system. Subsequently, the measurement is repeated and on success the state $|\downarrow_A\downarrow_B\rangle$ ($|\uparrow_A\uparrow_B\rangle$ after the π-pulse) can be removed, leaving the system in the maximally entangled Bell state

$$|\Psi^{\pm}\rangle = 1/\sqrt{2}\,(|\uparrow_A\downarrow_B\rangle \pm |\downarrow_A\uparrow_B\rangle), \quad (15.2)$$

where the sign depends on whether the two photons were observed on the same detector (+) or different detectors (-), respectively.

While in case of unity collection and detection efficiency $\eta = 1$ the success probability is $1/2$, it scales with η^2 in case of $\eta < 1$. A reduced detection efficiency reduces only the success probability and *not* the fidelity of the entangled state, when heralded properly. This is the key advantage of the scheme, making it intrinsically robust against photon losses. Nevertheless, a high detection and collection efficiency is desirable to increase the success probability.

15.1.2 *Heralded Entanglement with NV Centers*

The NV center in high-quality bulk diamond provides almost all prerequisites to apply the discussed heralded entanglement scheme. As discussed in Chapter 7, the ground state spin can be initialized, coherently controlled, and read out. Furthermore, with the spin-preserving transition between $m_s = 0$ ground state and $E_{x/y}$ excited state, the required L-type level scheme is present.

However, in the NV center, several hurdles have to be overcome. First, owing to the low Debye–Waller factor, the detection rate of indistinguishable photons from this transition is comparably low. This results in a low rate of successful entanglements, even when solid immersion lenses are used to improve the collection efficiency. Second, different NV centers exhibit slightly different transition frequencies due to strain. These have to be compensated by DC Stark shifts with electric fields. Third, the NV^- might be photo-ionized to NV^0, which has to be compensated by applying suitable repumping pulses. Fourth, the optical transitions are unstable due to spectral diffusion. To overcome this, the experiment must be conditioned on the position of the transition line or the transition must be stabilized actively. Fifth, the optical π-pulse must be separated from the subsequently emitted single photon. Here, spatial and temporal filters must be applied to reduce the contribution of scatter laser photons.

In a recent experiment all these hurdles are overcome, enabling the entanglement of two individual NV centers sitting in two cryostats separated by about 3 m [123]. In the experiment, the repetition rate of the entanglement operation is 20 kHz, while the success probability is only 10^{-7}, resulting in one heralded entanglement event per 10 minutes. The achieved overlap of the generated state ρ with the desired Bell state $\Psi^\pm$, i.e., the fidelity $F = \langle \Psi^\pm | \rho | \Psi^\pm \rangle$ is on the order of $F = 65\%$, clearly above the classical limit of $F = 50\%$.

In the current experiments the single photons are collected with solid immersion lenses allowing only low-success probabilities. Here, cavity-coupled systems might improve not only the collection efficiency but also the Debye–Waller factor. In particular, photonic crystal cavities seem promising, as all optical components and potentially also the photon detectors [378] might be integrated on one and the same chip, as illustrated in Fig. 15.2. When the entanglement rate can be successfully increased to the nuclear spin coherence time, the entanglement might get quasi-deterministic as discussed above. Then such devices might be scalable to many qubit

Figure 15.2 Artist's view of an integrated entanglement experiment. The two NV centers are located each in a L3 photonic crystal cavity, coupled to single-mode waveguides and an integrated 50/50 beams plitter (central region). Combined with integrated photon-routing mechanisms, this allows integrated larger-scale devices for quantum information processing.

systems, forming a promising quantum information processing platform. Apart from these promises, spectral diffusion is a serious hurdle which has to be overcome, independent of whether NV centers in diamond cavities or hybrid devices are used. Thus, alternative schemes that are less sensitive to fluctuations may render better suited. One promising example is the cavity-mediated deterministic entanglement scheme discussed in the following section.

15.2 Deterministic Entanglement

As discussed in the introduction of this chapter, a medium-range interaction on the order of a wavelength is required for fast operations between individual registers of a quantum node. Here, the probabilistic scheme is disadvantageous as it does not allow the required high-speed interactions.

Recently Yang et al. proposed that an effective interaction between medium-distance NVs in high-quality cavities with Q factors exceeding 10^6–10^8 is possible [356–359]. Achieving such high Q

factors in cavities with incorporated NV centers is technologically extremely challenging. In contrast, in this section the experimentally feasible case of photonic crystal cavities with Q factors of about 10^4–10^5 is regarded. It is shown that in this case efficient entanglement of medium-distance NV centers is possible using a scheme proposed for the first time by Imamoglu et al. in Ref. [379]. Although particularly photonic crystal cavities and NV centers are regarded, the scheme is applicable to other types of cavities and quantum systems.

In the following, first the model system is introduced and the analytical results of Ref. [379] are reviewed. Then, this model is adapted to realistic NV centers and a parameter range that has already been achieved in current experiments. The equations of motion are numerically solved, showing that the scheme can compete with other entanglement methods.

15.2.1 *The Model System*

The key elements of the entanglement scheme are two Λ-type systems (e.g., NV centers) with long-lived spin ground states $|0\rangle$, $|1\rangle$, in which a qubit can be encoded, and an excited state $|E\rangle$. These are placed in two antinodes of the mode of a low-Q photonic crystal cavity with small mode volume (Fig. 15.3). This configuration allows

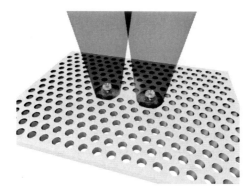

Figure 15.3 Artist's view of the considered system of two NV centers in nanodiamonds coupled via a photonic crystal cavity formed by a row of missing holes. From Ref. [377].

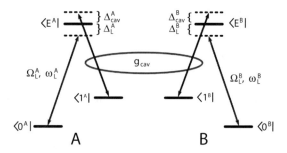

Figure 15.4 Level scheme of the two NV centers A and B. Each center provides a Λ scheme with ground states $|0^{A(B)}\rangle$, $|1^{A(B)}\rangle$ and excited state $|E^{A(B)}\rangle$. The $|0^{A(B)}\rangle \to |E^{A(B)}\rangle$ transitions are driven by lasers with frequency $\omega_L^{A(B)}$ and coupling strength $\Omega_L^{A(B)}$ detuned by $\Delta_L^{A(B)}$. The $|1^{A(B)}\rangle \to |E^{A(B)}\rangle$ transition of each system is coupled to the shared cavity mode with coupling strength $g_{cav}^{A(B)}$, where the cavity is detuned by $\Delta_{cav}^{A(B)}$. From Ref. [377].

for independent optical access to both systems, enabling established optical schemes for initialization and readout. Furthermore, this configuration allows coherent all-optical one-qubit operations, e.g., in the Raman scheme (Section 4.3): Two laser fields, almost resonant with the optical transition, with coupling constant $\Omega_{L(R)}$, frequency $\omega_{L(R)}$, and frequency difference $\delta\omega = \omega_L - \omega_R$ corresponding to the energy spacing ω_{01} between $|0\rangle$ and $|1\rangle$ are applied to an individual system. If the lasers are detuned by Δ_L from the transition to the excited state, the system undergoes a spin rotation with the frequency $\Omega_{Raman} = \Omega_L \Omega_R / (2\Delta_L)$. Using this Raman scheme, any single qubit operation can be performed.

A universal two-qubit operation is the spin exchange [379]. For this, one of the Raman lasers is applied to each system, while the second laser is replaced by the cavity mode, as depicted in Fig. 15.4. Importantly, the cavity is detuned by $\Delta_{cav}^{A(B)}$ from the Raman resonance condition used in the conventional Raman scheme. Now, both system are simultaneously driven with laser fields, and a coherent spin-exchange by stimulated Raman scattering takes place: For example system A, initially prepared in $|0\rangle$, emits a Raman photon into the cavity mode, while undergoing a spin flip. This process is virtual and can only occur within the time–energy uncertainty, as the photon frequency does not match the cavity resonance. Only if the photon is absorbed in a second Raman process, where system B undergoes a spin flip, the energy is conserved and the joint spin flip process occurs.

15.2.2 *Effective Hamiltonian Approach*

To quantify this, the system of $|1\rangle$, $|0\rangle$, and $|E\rangle$ is described by the Hamiltonian $H = H_0 + H_I$. When, for simplicity, assuming equal parameters for both λ systems, the free H_0 and interaction part H_I read in the rotating wave approximation

$$H_0 = \sum_{i,j} \hbar \omega_i |i_j\rangle \langle i_j| + \hbar \omega_{cav} c^\dagger c, \tag{15.3}$$

$$H_I = \sum_j \hbar \left[\Omega^j |0_j\rangle \langle E_j| e^{i\omega_L t} + g_{cav}^j c^\dagger |1_j\rangle \langle E_j| \right] + \text{h.c.,} \tag{15.4}$$

where $i \in \{0, 1, E\}$, $j \in \{A, B\}$, g_{cav} denotes the cavity coupling for the $|1\rangle \leftrightarrow |E\rangle$ transition and $c, c^\dagger$ are the usual operators for cavity photons.

When the decay constant of the cavity mode $\kappa = \omega_{cav}/Q$ is small compared to $\omega_{10} = \omega_1 - \omega_0$ and the cavity detuning Δ_{cav}, i.e., the Q factor is very high, one can show analytically [379] that an effective interaction between the two spins is generated. This interaction has the form

$$H_{eff} = \frac{1}{2} \tilde{g} e^{i\Delta_{AB}t} |0_A\rangle \langle 1_A| \otimes |1_B\rangle \langle 0_B| + \text{h.c.,} \tag{15.5}$$

with the effective overall detuning from the virtual Raman resonance $\Delta_{AB} = \Delta_L^A + \Delta_{cav}^A - \Delta_L^B - \Delta_{cav}^B$ and the effective coupling constant $\tilde{g} = g_{eff}^A g_{eff}^B / (\Delta_L^A - \Delta_{cav}^A)$, where

$$g_{eff}^{A(B)} = \frac{g_{cav}^{A(B)} \Omega^{A(B)}}{2(\Delta_L^{A(B)} + \Delta_{cav}^{A(B)})}. \tag{15.6}$$

The time evolution described by the Hamiltonian Eq. (15.5) is an effective rotation $U_{exc}(\varphi)$ on the two-spin state $|S_A, S_B\rangle$. To generate an entangled state, system A is prepared in the state $|0_A\rangle$ while system B is prepared in the state $|1_B\rangle$, i.e., the system of two spins is prepared in the state $|01\rangle$. Now, by applying a $U_{exc}(\pi/2)$ spin-exchange, this state is transformed into $|\Psi\rangle = 1/\sqrt{2}(|01\rangle + i|10\rangle)$, a maximally entangled state. This entanglement operation (EO) has several important properties:

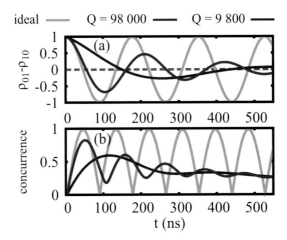

Figure 15.5 Dynamics of the two NV spins. (a) Calculated inversion ρ_{01} − ρ_{10} between initial state and target state for the ideal case ($\kappa = \gamma = 0$) and different Q factors. Even with very moderate Q factors, significant population transfer is possible. (b) Calculated concurrence for (a) indicating the generation of an entangled state during the transfer. Even for a Q factor as low as 10^4, a high concurrence can be achieved. For all calculations, $\Delta_{cav} = 9g_{cav} + 2\kappa$, $\Delta_L = 9g_{cav}$, $\Omega = g_{cav} = 2\pi \cdot 3$ GHz is used. From Ref. [377].

(1) If the detunings Δ_{cav} and Δ_L are large enough, the mechanism is robust against spectral diffusion, i.e., small fluctuations of the excited state.

(2) It is not necessary that the two systems are equal. Differences in the optical transition frequency can be compensated by a proper choice of laser frequencies.

(3) Emitters that are detuned from the resonance, i.e., $\Delta_{AB} \neq 0$ or outside the laser focus, are unaffected, making the mechanism scalable to several systems inside one single cavity.

(4) By applying single qubit unitary transformations and several spin exchanges, the fundamental c-Not gate can be constructed [379].

In the following, it is analyzed how these results are applicable to experimentally feasible implementations, regarding in particular the losses from the cavity and radiative dephasing of the excited state $|E\rangle$.

15.2.3 Lindblad Approach

Including dissipative processes, the equation of motion for the density matrix ρ is given by $d\rho/dt = -i/\hbar[\rho, H]_- + \mathcal{L}(\rho)$, with the Lindblad form

$$\mathcal{L} = \sum_x \hat{\gamma}_x \rho \hat{\gamma}_x^\dagger - \frac{1}{2}[\hat{\gamma}_x^\dagger \hat{\gamma}_x, \rho]_+ + \hat{\kappa}\rho\hat{\kappa}^\dagger - \frac{1}{2}[\hat{\kappa}^\dagger\hat{\kappa}, \rho]_+. \quad (15.7)$$

Here, $x \in \{0^A, 0^B, 1^A, 1^B\}$, $\hat{\gamma}_x = \sqrt{\gamma}|x\rangle\langle E|$, with $\gamma = 50$ MHz describing the decay from the exited stated to ground state x under emission into non-cavity modes and $\hat{\kappa} = \sqrt{\kappa}c$ losses from the cavity.

This equation of motion for the components of the density matrix is expanded and solved using an explicit Runge–Kutta algorithm.

For the cavity coupling $g_{cav}/(2\pi) = 3.0$ GHz is chosen for NV centers sitting in the field maximum of a nanocavity [360]. This is feasible by slightly improving experimental results on the Purcell enhancement of the zero-phonon transition of NV centers in photonic crystal L3 cavities [128, 319]: With $F = 12\,(60)$ being the demonstrated Purcell factor [128] ([319]), $Q = 600\,(3000)$ the quality factor of the used cavity, $\tau = 14$ ns the lifetime of the excited state, $d = 0.05$ the Debye–Waller factor, and $\omega/(2\pi) = 671$ THz the frequency of the optical NV transition, the experimentally achieved coupling is calculated to

$$\frac{g_{cav}}{2\pi} = \frac{1}{2\pi}\sqrt{\frac{d\omega F}{4Q\tau}} = 1.15\,\text{GHz}. \quad (15.8)$$

Here it is assumed that $\Omega = g_{cav}$, which can be achieved even for spin nonpreserving transitions with laser powers of about 1 mW [124, 380]. The laser detuning is set twice the cavity coupling $\Delta_L^0 = 9g_{cav}$, while the cavity detuning is chosen according to $\Delta_{cav}^0 = 9g_{cav} + 2\kappa$. These values represent a good compromise between radiative dephasing, cavity losses, and the time needed for the EO. Furthermore, without loss of generality, the ground state splitting is set to the zero field splitting of $\omega_{12} = 2\pi \cdot 2.87$ GHz. With these parameters the dynamics for $Q = \omega/\kappa = 9800$ are calculated, which is in the range of current experiments.

Starting with NV^A in state $|0\rangle^A$ and NV^B in state $|1\rangle^B$, i.e., the diagonal elements $\rho_{01} = 1$, $\rho_{00} = \rho_{10} = \rho_{11} = 0$ a spin exchange takes place, as predicted by the analytical theory. The maximally achieved inversion is $\rho_{01} - \rho_{10} > 0.3$, after a transfer time of about 300 ns (Fig. 15.5). To confirm that the transfer is indeed coherent and an entangled state is prepared, the *concurrence c* [381] as a positive definite measure of entanglement during the transfer is evaluated. A vanishing concurrence indicates a classical, i.e., separable state, while a concurrence of 1 indicates a maximally entangled state. Even with the low $Q = 9800$, a value of $c_{max} \approx 0.6$ is found for the maximally achieved concurrence after the time $t_{max} \approx 150$ ns. This strikingly demonstrated that even low-Q photonic crystal cavities can mediate entanglement between two NV centers. using the challenging but nevertheless realistic value of $Q = 98000$, the EO gets even improved. In this case, a maximal inversion of $\rho_{01} - \rho_{10} > 0.6$ and a maximal concurrence of $c_{max} \approx 0.8$ are found.

15.2.4 *Influence of the Detunings and Spectral Diffusion*

To study the robustness of the scheme against small fluctuations of the cavity, the laser, and the NV's optical transition frequency ω [154], the influence of the detunings Δ_L, Δ_{cav} on the entanglement generation is calculated (Fig. 15.6). In the parameter regime studied here, the analytical theory suggests a *linear* dependency between the EO time t_{max} and these detunings, i.e., $t_{max} \sim 1/\tilde{g} \sim \Delta_L$, Δ_{cav}. Actually, the numerical solution confirms this prediction for the cavity detuning Δ_{cav}^0. An increasing Δ_{cav} gives rise to a linear increased entanglement time t_{max} [Fig. 15.6(b)]. Here, the initial choice of Δ_{cav}^0 represents indeed a good compromise between the efficiency of the EO [Fig. 15.6(a)] and the entanglement time. Similarly, the initially chosen cavity detuning Δ_L^0 is close to its optimal value, allowing for high concurrence at low EO times. Surprisingly, the numerically calculated EO time depends almost *quadratically* on the laser detuning Δ_L. This is in clear disagreement with the analytical theory for $1/\tilde{g}$, where both detunings contribute equally and linearly.

When the frequency of the optical transition changes by $\Delta\omega$ due to spectral diffusion, this is equivalent to a simultaneous change of the laser detuning by $\pm\Delta\omega$ and the cavity detuning by $\mp\Delta\omega$. Here, the opposite signs guarantee robustness of the EO against spectral

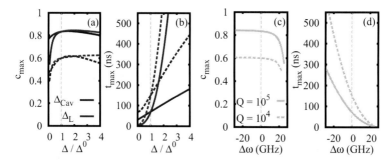

Figure 15.6 Influence of the laser and cavity detuning Δ_L, Δ_{cav} on the entanglement generation. (a) The achieved maximum concurrence for different Q factors when varying Δ_L or Δ_{cav}, while keeping the respective other detuning fixed at Δ^0. For the calculations it is assumed that $g_{cav} = \Omega = 2\pi \cdot 3$ GHz, $\Delta_L^0 = 9g_{cav}$, $\Delta_{cav}^0 = 9g_{cav} + 2\kappa$, and $Q = 10^4$ (dashed lines), or $Q = 10^5$ (solid lines), respectively. (b) The transfer time needed to achieve the concurrence in plotted (a). Achievable concurrence (c) and EO time (d), when the optical transition frequency changes by $\Delta\omega$. While the concurrence is robust even for fluctuations as large as 10 GHz, spectral diffusion might lead to dephasing, as the EO time t_{max} changes by about 7% per GHz shift of the optical transition. From Ref. [377].

diffusion. Indeed, the achievable concurrence is almost invariant for $\Delta\omega < 10$ GHz [Fig. 15.6(c)]. Nevertheless, the EO times changes by about 7% per GHz shift and thus dephasing might occur when spectral diffusion exceeds a few gigahertz [Fig. 15.6(d)].

15.2.5 *Influence of Q factor and Cavity Coupling*

To investigate the influence of the cavity quality factor Q and coupling g_{cav} in detail, the maximum achievable concurrence c is shown in Fig. 15.7(a) and the needed entanglement time t_{max} is shown in Fig. 15.7(b) as a function of Q for various couplings between $g_{cav} = 2\pi \cdot 0.3$ GHz and $g_{cav} = 2\pi \cdot 3.0$ GHz. As expected, for small Q factors, photon loss from the cavity modes limits the achievable concurrence. On the other hand, a strong dependency on the coupling constant g_{cav} is visible. This can be explained by Raman scattering into non-cavity modes that induces additional unintended spin flips and dominates the dynamics for low ratios between g_{cav} and γ.

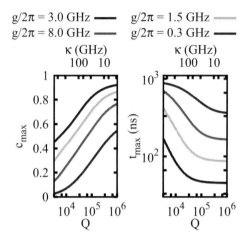

Figure 15.7 Entanglement generation as a function of the Q factor. (a) The achieved concurrence for different cavity couplings g_{cav} between $2\pi \cdot 3.0\,\text{GHz}$ and $2\pi \cdot 0.3\,\text{GHz}$ when varying the Q factor. Even with moderate Q factors of 10^4 an entangled state can be prepared. (b) The transfer time needed to achieve the concurrence in (a). For all calculations, $\Delta_{cav} = 8g_{cav} + 2\kappa$, $\Delta_L = 8g_{cav}$, $\Omega = g_{cav}$ is used. From Ref. [377].

In conclusion, small mode volume photonic crystal cavities with comparably low Q factors can be an important tool on the path towards deterministic entanglement of medium-distance NV centers. This might be a key ingredient for future quantum information processing networks under realistic conditions, including fluctuations and dephasing. Thus, in the field of integrated quantum hybrid systems, many exciting experiments are expected in the future.

Chapter 16

Conclusions and Outlook

The subject of the present work was the rich physics and demanding technology of integrated quantum hybrid systems. Such systems, built from classical dielectric nanostructures and individual quantum objects, promise to provide a scalable platform to exploit the laws of quantum physics for efficient computation, secure communication, and measurements with unreached accuracy. Ultimately, they might even serve to explore the limits of the formalism of quantum mechanics. Hence, integrated quantum-optical hybrid systems are in the focus of many researchers worldwide.

16.1 Summary and Conclusions

To obtain an instructive introduction into the topic, the text was divided into four parts, which are summarized one by one in the following. The references in the summaries of Part I–IV exclusively refer to results obtained by the author.

Part I

To provide an insight into the topic, Part I was devoted to the fundamental theoretical concepts of optical active quantum systems

Integrated Quantum Hybrid Systems
Janik Wolters
Copyright © 2015 Pan Stanford Publishing Pte. Ltd.
ISBN 978-981-4463-82-9 (Hardcover), 978-981-4463-83-6 (eBook)
www.panstanford.com

and quantum electrodynamics. Here, the electromagnetic field and its quanta, the photons, were introduced and their fundamental properties were reviewed. Later on, different approaches to light–matter interaction were provided. The incoherent dynamics of two- and three-level systems were discussed in second-order perturbation theory, i.e., using Fermi's golden rule. This made it possible to describe fundamental absorption and emission processes and to predict the statistical behavior of photons emitted by individual quantum systems. The coherent dynamics were discussed in the semiclassical approach, on the basis of the optical Bloch equations and their extension to three-level systems. Here, the fundamental Rabi oscillations and the Bloch sphere were introduced. The effect of power broadening was found and the influence and origin of decoherence was treated. Furthermore, the Bloch equations were extended to Λ-type three-level systems, and stimulated Raman transitions were studied. At the end of Part I, cavity modes and the fully quantized Jaynes–Cummings model were introduced. From this model the single-photon Bloch equations were derived and fundamental effects like vacuum Rabi splitting, vacuum Rabi oscillations, and the Purcell effect were described. These effects outlined how resonator structures can be used to tailor the light–matter interaction on the level of individual quanta.

Part II

The second part of the present work was dedicated to the different solid state systems, which are suitable for integration into quantum hybrid devices. Here, techniques to investigate individual quantum systems and the experimental results were discussed. At first, the fundamental physics of quantum dots was reviewed and experiments were presented. So, on-demand single-photon emission and the generation of entangled photon pairs via the biexciton cascade could be achieved. Additionally, ultrafast coherent control of the electron spin in charged quantum dots could be demonstrated, although the short coherence time appears as a major drawback. Furthermore, optically active organic molecules were introduced. Using such molecules, stable room temperature single-photon emission could be demonstrated, while more complex experiments using, e.g., electron spins were limited by the lack of optical active triplet states. In contrast, color centers in diamond were identified to overcome many of these problems. In particular, the negatively charged nitrogen-vacancy center provides an optically active triplet ground state, all-optical spin initialization and readout mechanisms, and extremely long coherence times. Here, not only single-photon generation, but also experiments on the coherent spin manipulation and its inhibition by the quantum Zeno effect [190] were presented. Nevertheless, spectral diffusion was identified as a major hurdle for future integrated devices that shall benefit from the controlled interaction of many color centers [154].

Part III

The second constituent of integrated hybrid devices, dielectric nano- and microstructures, were introduced in Part III. Starting with Maxwell's equations, the fundamental equations of electrodynamics in dielectric materials were elaborated. Here, modified expressions for the electric field per photon and spontaneous emission as well as the mechanism of total internal reflection were derived. At first glance, total internal reflection leads to the problem of photon extraction and the treatment of immersion microscopy techniques. At the second glance, total internal reflection allowed light guiding and confinement in waveguides and microdisk resonators. As a second concept for confining light photonic bandgap materials, the photonic crystals were introduced. For these, fabrication issues, as well as experimental techniques to investigate photonic crystal resonators, were addressed. Part III was completed by a discussion of several classical applications of photonic crystals cavities. Recent experimental results on refractive index measurements in ultrasmall volumes and thermo-optical switching [252] were discussed.

Part IV

In the last part the previously discussed components, namely solid state quantum systems and dielectric structures, were combined to the actual quantum hybrid systems. Recently developed tools for building such devices were introduced, and landmark experimental results on the Purcell enhancement of single quantum dots and single nitrogen-vacancy centers in nanodiamond [128, 129] were discussed. These experimental results promised a feasible road toward the efficient generation of narrow-band single photons for future quantum communication, computation, and metrology. Subsequently, the strong coupling regime of cavity quantum electrodynamics was treated. It was found that this regime could be reached in recent experiments with quantum dots in microcavities. For nitrogen-vacancy centers this is not within the reach of today's technology, because of the comparable weak dipole moment. However, two entanglement schemes that do *not* require strong coupling were introduced. It was found that in a probabilistic scheme, as well as in a deterministic scheme, high-fidelity entanglement of medium-distance nitrogen-vacancy centers on a chip is feasible with current technology [377].

16.2 Outlook

This work gave an overview of the fundamentals and recent developments on quantum hybrid systems. Although tremendous progress was made in the recent years, integrated quantum hybrid technology is still in its infancy. Thus, many interesting experiments and applications are expected in the future.

The application that is closest to being market-ready is certainly quantum communication [382]. In quantum communication, single photons are used to carry information and allow communication exclusively secured by the laws of quantum mechanics. To achieve a high transmission bandwidth, ultrabright and efficient on-demand single-photon sources are required. Here, Purcell-enhanced single emitters, as discussed in Chapter 13, are certainly the photon source of choice. Once such integrated device, available on an industrial scale, will revolutionize the quantum communication market.

Owing to unavoidable transmission losses in optical fibers, the achievable communication distance with single photons is limited to the order of 100 km, i.e., metropolitan areas. For long-range or even intercontinental quantum communication, relay stations, so-called quantum repeaters, must be developed [74]. The most popular schemes for such quantum repeaters rely on entanglement generation and storage, and subsequent quantum teleportation for entanglement swapping. Hence, maintenance-free entanglement generating and storing devices are demanded. Here, the biexciton cascade (Section 5.2.2) and the entanglement schemes from Chapter 15 can be applied. Additionally, it is required that all repeater stations somehow transform the locally generated entanglement into indistinguishable photons, and vice versa. Thus, efficient matter–photon interfaces with equal emitters are required. Here, cavity-enhanced interactions will certainly be helpful.

When the quantum repeaters are available, their technology can also be used to construct quantum computers and simulators. In such quantum information-processing apparatuses, entanglement between many individual systems can be exploited to solve complex mathematical and physical problems. Here, the scalability of integrated solid state devices is a major advantage above a technology relying on single atoms or ions in the gas phase, which are intrinsically hard to scale. Once the quantum information-processing schemes are successfully implemented, the promises and prospects of an integrated quantum technology platform can be fully exploited. Then, they might even serve to explore the limits of our understanding of the physical world, i.e., the formalism of quantum mechanics.

Acknowledgments

The present text is a synopsis of the knowledge I gained and the work I did during three-and-a-half years in the Nano-Optics group at Humboldt-Universität zu Berlin. Of course, writing this monograph would have been impossible without the support of many people and institutions, whom Id like to thank here.

First of all I have to thank my PhD supervisor, Prof. Oliver Benson, for providing the framework of my work and his constant support and advice. In addition, I want to express my gratitude toward him and Prof. Jelena Vuckovic for contributiong the foreword of this book.

Then, I am very grateful to Günter Kewes, Niko Nikolay, Carlo Barth, Nikola Sadzak, Michael Adler, and Max Strauß, who supported my work while they were busy writing their respective bachelors, masters, and diploma theses. Thanks to Klaus Palis for his assistance, advice, and patience with many electronics projects. Furthermore, thanks to my fellow PhD students, in particular Andreas W. Schell, as well as to the postdocs from the group, in particular Thomas Aichele and Simon Schönfeld, for the many discussions I had with them and for their unfailing assistance in joint projects. It was a pleasure to work with them.

I am very happy about the fruitful, instructive, and interesting collaborations with Nils Nüsse, Max Schoengen, Jürgen Probst, Bernd Löchel, Henning Döscher, Thomas Hannappel, Helmut Fedder, Fedor Jelezko, Hannes Bernien, Ronald Hanson, Mladen Pavicic, Costanza Toninelli, Julia Kabuß, Andreas Knorr, Oliver Schöps, Nicolai Grosse, and Ulrike Woggon. Without these people, a major part of my work would have been impossible.

I have to thank the DFG and the State of Berlin for funding me and my projects, as well as Pan Stanford Publishing and PicoQuant GmbH for their collaboration and support.

Last but not the least, I am very pleased with the patience of Linda and Anton.

Janik Wolters
Berlin
June 2015

Own Contributions

Peer Review Articles and Book Chapters

Partially, this work is based on previous peer review publications and book chapters by the author. These self-citations are summarized below in chronological order.

Section 15.2

"Deterministic and robust entanglement of nitrogen-vacancy centers using low-Q photonic-crystal cavities," **J. Wolters**, J. Kabuß, A. Knorr and O. Benson, *Physical Review A* **89**, 060303 (2014).

Introduction of Chapter 6 and Section 7.1

"Engineering and applications of fundamental photonic elements based on defect centres in nanodiamonds," A. W. Schell, **J. Wolters**, T. Schröder, and O. Benson in *Quantum information processing with diamond*, S. Prawer and I. Aharonovich (Eds.), Woodhead Publishing (2014).

Sections 4.2.9, 7.6, 7.7, and 7.5.2

"Observation of the quantum Zeno effect on a single solid state spin," **J. Wolters**, M. Strauß, R. S. Schoenfeld, and O. Benson, *Physical Review A* **88**, 020101 (2013) [190].

Section 7.4

"Measurement of the ultrafast spectral diffusion of the optical transition of nitrogen vacancy centers in nano-size diamond using correlation interferometry," **J. Wolters**, N. Sadzak, A. W. Schell, T. Schröder, and O. Benson, *Physical Review Letters* **110**, 027401 (2013) [154].

Sections 10.3.1, 10.3.2, and 13.2.1

"Three-dimensional quantum photonic elements based on single nitrogen vacancy-centres in laser-written microstructures," A. W. Schell, J. Kaschke, J. Fischer, R. Henze, **J. Wolters**, M. Wegener, and O. Benson, *Scientific Reports* **3**, 1577 (2013) [198].

Sections 11.4.1, 11.5.2, 12.2, and 12.3

"Thermo-optical response of photonic crystal cavities operating in the visible spectral range," **J. Wolters**, N. Nikolay, M. Schoengen, A. W. Schell, J. Probst, B. Löchel, and O. Benson, *Nanotechnology* **24**, 315204 (2013) [252].

Sections 11.6 and 13.2.2

"Coupling of single nitrogen-vacancy defect centers in diamond nanocrystals to optical antennas and photonic crystal cavities," **J. Wolters**, G. Kewes, A. W. Schell, N. Nüsse, M. Schoengen, B. Löchel, T. Hanke, R. Bratschitsch, A. Leitenstorfer, T. Aichele, and O. Benson, *Physica Status Solidi (B)* **249**, 918 (2012) [129].

Section 13.2.2

"A scanning probe-based pick-and-place procedure for assembly of integrated quantum optical hybrid devices," A. W. Schell, G. Kewes, T. Schröder, **J. Wolters**, T. Aichele, and O. Benson, *Review of Scientific Instruments* **82**, 073709 (2011) [330].

Section 13.2.2

"Enhancement of the zero phonon line emission from a single nitrogen vacancy center in a nanodiamond via coupling to a photonic crystal cavity," **J. Wolters**, A. W. Schell, G. Kewes, N. Nüsse,

M. Schoengen, H. Döscher, T. Hannappel, B. Löchel, M. Barth, and O. Benson, *Applied Physics Letters* **97**, 141108 (2010) [128].

Further publications

Further publications that influenced this work, but are not cited directly are (in chronological order):

"Entanglement of two ground state neutral atoms using Rydberg blockade," Y. Miroshnychenko, A. Browaeys, C. Evellin, A. Gaëtan, T. Wilk, **J. Wolters**, P. Grangier, A. Chotia, D. Comparat, P. Pillet, and M. Viteau, *Optics and Spectroscopy* **111**, 540 (2011).

"Analysis of the entanglement between two individual atoms using global Raman rotations," A. Gaëtan, C. Evellin, **J. Wolters**, P. Grangier, T. Wilk, and A. Browaeys, *New Journal of Physics* **12**, 065040 (2010).

"Entanglement of two individual neutral atoms using Rydberg blockade," T. Wilk, A. Gaëtan, C. Evellin, **J. Wolters**, Y. Miroshnychenko, P. Grangier, and A. Browaeys, *Physical Review Letters* **104**, 010502 (2010).

"Theory of carrier and photon dynamics in quantum dot light emitters," M.-R. Dachner, E. Malic, M. Richter, A. Carmele, J. Kabuss, A. Wilms, J. Kim, G. Hartmann, **J. Wolters**, U. Bandelow, and A. Knorr, *Physica Status Solidi (b)* **247**, 809 (2010).

"Carrier heating in light-emitting quantum-dot heterostructures at low injection currents," **J. Wolters**, M.-R. Dachner, E. Malic, M. Richter, U. Woggon, and A. Knorr, *Physical Review B* **80**, 245401 (2009).

Conference Proceedings

"Nanophotonics with single photons from NV centers in three-dimensional laser-written microstructures," A. W. Schell, T. Neumer, Q. Shi, J. Kaschke, J. Fischer, R. Henze, **J. Wolters**, M. Wegener, and O. Benson, *Frontiers in Optics 2013, FW1C.2* (2013).

"Revealing spectral diffusion in single nitrogen-vacancy centers in nanodiamond by photon correlation interferometry," N. Sadzak, J. **Wolters**, A. W. Schell, T. Schrögder, O. Benson, *Frontiers in Optics 2013, FTu1C.3* (2013).

"Observation of the quantum Zeno effect on the nitrogen vacancy center in nanodiamond," J. **Wolters**, M. Strauß, R. S. Schönfeld, and O. Benson, *CLEO: 2013, OSA Technical Digest*, QM2B.5 (2013).

"Single photon nanophotonics using NV centers in three-dimensional laser-written microstructures" A. W. Schell, J. Kaschke, J. Fischer, R. Henze, J. **Wolters**, M. Wegener, and O. Benson, *European Conference on Lasers and Electro-Optics and XIIIth International Quantum Electronics Conference, OSA Technical Digest*, CK-7.1 (2013).

"Demonstration of the quantum Zeno effect on the nitrogen vacancy center in nanodiamond," J. **Wolters**, M. Strauß, R. S. Schönfeld, and O. Benson, *European Conference on Lasers and Electro-Optics and XIIIth International Quantum Electronics Conference, OSA Technical Digest*, IA-3.2 (2013).

"Nanodiamonds for integrated quantum technology: charm and challenge," J. **Wolters**, A. W. Schell, N. Sadzak, T. Schröder, M. Schoengen, J. Probst, B. Löchel, and O. Benson, *Research in Optical Sciences, OSA Technical Digest*, QW1B.3 (2012).

"Assembly of quantum optical hybrid devices via a scanning probe pick-and-place technique," A. W. Schell, J. **Wolters**, G. Kewes, T. Schröder, T. Aichele, and O. Benson, *CLEO: QELS-Fundamental Science, OSA Technical Digest*, QW3H.2 (2012).

"Quantum light emission from cavity enhanced LEDs," A. Carmele, M.-R. Dachner, J. **Wolters**, M. Richter, and A. Knorr, *10th International Conference on Numerical Simulation of Optoelectronic Devices (NUSOD), 2010*, 85 (2010).

"Theory of few photon dynamics in electrically pumped light emitting quantum dot devices," A. Carmele, M.-R. Dachner, J. **Wolters**, M. Richter, and A. Knorr, *Proc. SPIE* **7597**, 75971 (2010).

"Assembly of fundamental photonic elements from single nanodi-amonds", T. Aichele, A. Schell, M. Barth, S. Schietinger, T. Schröder,

J. Wolters, O. Benson, N. Nüsse, and B. Löchel, *2010 IEEE Photonic Societys 23rd Annual Meeting*, 142 (2010).

"Coupled carrier-phonon dynamics in light emitting quantum-dot heterostructures: switch on dynamics and carrier heating," **J. Wolters**, M. R. Dachner, M. Richter, A. Knorr, and U. Woggon, *Conference on Lasers and Electro-Optics/International Quantum Electronics Conference, OSA Technical Digest*, JTuD108 (2009).

"Effective Hamiltonian approach to multiphonon effects in self assembled quantum dots," M. Dachner, **J. Wolters**, A. Knorr, and M. Richter, *Conference on Lasers and Electro-Optics/International Quantum Electronics Conference, OSA Technical Digest*, JWA119 (2009).

"Theory of electron dynamics in light emitting quantum dot devices," M. Richter, E. Malic, J. Kim, M. R. Dachner, **J. Wolters**, U. Woggon, and A. Knorr, *CLEO/Europe and EQEC 2009 Conference Digest*, CB14.6 (2009).

Other Conference Contributions

Invited Talks

- Workshop on Quantum Information using NV Centres in Diamond, Bonamanzi, South Africa (2012).
- MRS Spring Meeting, San Francisco, USA (2014).

Contributed Talks

- DPG Spring Meeting, Dresden, Germany (2011).
- DPG Spring Meeting, Stuttgart, Germany (2012).
- MRS Spring Meeting, San Francisco, USA (2012).
- International Conference on Diamond and Carbon Materials, Granada, Spain (2012).
- DPG Spring Meeting, Regensburg, Germany (2013).
- DPG Spring Meeting, Hannover, Germany (2013).
- International Conference on Quantum Information Processing and Communication, Florence, Italy (2013).
- International Conference on Diamond and Carbon Materials, Riva del Garda, Italy (2013).

Theses

Own Diploma Thesis

"Self consistent theory of nonequilibrium phonons in semiconductor devices," Janik Wolters, Technische Universität Berlin, Germany (2009).

Supervised Master and Diploma Theses

"Quantum optics experiments with single nitrogen-vacancy centers in nanodiamonds," Nikola Sadzak, Universita degli Studi di Padova, Italy (2012).

"Simulation, fabrication and characterization of photonic crystals," Michael Adler, Technische Universität Berlin, Germany (2013).

"Coherent spin manipultation on single nitrogen vacancy centers," Max Strauß, Humboldt-Universität zu Berlin, Germany (provisioned 2013).

"Investigation of grating-couplers for photonic crystal structures," Carlo Barth, Humboldt-Universität zu Berlin, Germany (provisioned 2014).

"Investigation of components for integrated quantum devices based on nitrogen vacancy centers," Niko Nikolay, (provisioned 2014).

"Coherent optical control of nitrogen vacancy centers," Bernd Sontheimer, Humboldt-Universität zu Berlin, Germany (provisioned 2014).

Supervised Bachelor Theses

"Design and optimization of waveguides and grating-couplers for photonic crystal structures," Carlo Barth, Humboldt-Universität zu Berlin, Germany (2011).

"Investigation of the temperature dependency of resonances in photonic crystal structures," Niko Nikolay, Humboldt-Universität zu Berlin, Germany (2011).

References

1. M. Polanyi. The stability of beliefs. *The British Journal for the Philosophy of Science* III, 217 (1952).

2. H. Steinwedel and W. Paul. Apparatus for separating charged particles of different specific charges, US Patent No. US 2939952 (1953).

3. F. Diedrich and H. Walther. Nonclassical radiation of a single stored ion. *Physical Review Letters* **58**, 203 (1987).

4. W. Moerner and L. Kador. Optical detection and spectroscopy of single molecules in a solid. *Physical Review Letters* **62**, 2535 (1989).

5. P. Michler, A. Imamoglu, M. D. Mason, P. J. Carson, G. F. Strouse, and S. K. Buratto. Quantum correlation among photons from a single quantum dot at room temperature. *Nature* **406**, 968 (2000).

6. C. Kurtsiefer, S. Mayer, P. Zarda, and H. Weinfurter. Stable solid-state source of single photons. *Physical Review Letters* **85**, 290 (2000).

7. N. Schlosser, G. Reymond, I. Protsenko, and P. Grangier. Sub-poissonian loading of single atoms in a microscopic dipole trap. *Nature* **411**, 1024 (2001).

8. J. Bardeen and W. Brattain. The Transistor, A Semi-Conductor Triode. *Physical Review* **74**, 230 (1948).

9. K. C. Kao and G. A. Hockham. Dielectric-fibre surface waveguides for optical frequencies. *Proceedings of the Institution of Electrical Engineers* **113**, 1151 (1966).

10. T. Monz, P. Schindler, J. T. Barreiro, M. Chwalla, D. Nigg, W. A. Coish, M. Harlander, W. Hänsel, M. Hennrich, and R. Blatt. 14-Qubit entanglement: creation and coherence. *Physical Review Letters* **106**, 130506 (2011).

11. G. Grynberg, A. Aspect, and C. Fabre. *Introduction to Quantum Optics: From the Semi-classical Approach to Quantized Light.* Cambridge University Press (2010).

12. J. D. Jackson. Classical Electrodynamics. John Wiley & Sons, Inc. (1999).

13. C. Gerry and P. Knight. *Introductory Quantum Optics*. Cambridge University Press (2004).

14. M. Kira, F. Jahnke, W. Hoyer, and S. W. Koch. Quantum theory of spontaneous emission and coherent effects in semiconductor microstructures. *Progress in Quantum Electronics* **23**, 189 (1999).

15. E. Schrödinger. Quantisierung als Eigenwertproblem. *Annalen der Physik* **384**, 361 (1926).

16. W. Heisenberg. Über quantentheoretische Umdeutung kinematischer und mechanischer Beziehungen. *Zeitschrift für Physik* **33**, 879 (1925).

17. P. A. M. Dirac. The quantum theory of the emission and absorption of radiation. *Proceedings of the Royal Society A: Mathematical, Physical and Engineering Sciences* **114**, 243 (1927).

18. H. Haken. *Quantum Field Theory of Solids: An Introduction*. Elsevier Science & Technology Books (1976).

19. J. J. J. Sakurai and J. Napolitano. *Modern Quantum Mechanics*. Addison Wesley. (2011).

20. H. Paul. *Photonen*. Teubner Studienbücher Physik. Vieweg+Teubner Verlag (1999).

21. W. Nolting. *Grundkurs Theoretische Physik 5/1: Quantenmechanik – Grundlagen*. Springer-Lehrbuch. Springer London Ltd. (2007).

22. P. Grangier, J. A. Levenson, and J.-P. Poizat. Quantum non-demolition measurements in optics. *Nature* **396**, 175 (1998).

23. C. Guerlin, J. Bernu, S. Deléglise, C. Sayrin, S. Gleyzes, S. Kuhr, M. Brune, J.-M. Raimond, and S. Haroche. Progressive field-state collapse and quantum non-demolition photon counting. *Nature* **448**, 889 (2007).

24. R. H. Hadfield. Single-photon detectors for optical quantum information applications. *Nature Photonics* **3**, 696 (2009).

25. R. Glauber. The quantum theory of optical coherence. *Physical Review* **130**, 2529 (1963).

26. G. A. Steudle, S. Schietinger, D. Höckel, S. N. Dorenbos, I. E. Zadeh, V. Zwiller, and O. Benson. Measuring the quantum nature of light with a single source and a single detector. *Physical Review A* **86**, 053814 (2012).

27. R. Q. Hanbury Brown and R. Twiss. Correlation between Photons in two Coherent Beams of Light. *Nature* **177**, 27 (1956).

28. F. Jelezko and J. Wrachtrup. Single defect centres in diamond: a review. *Physica Status Solidi (A)* **203**, 3207 (2006).

29. F. Bloch. Nuclear Induction. *Physical Review* **70**, 460 (1946).

30. M. Citron, H. Gray, C. Gabel, and C. Stroud. Experimental study of power broadening in a two-level atom. *Physical Review A* **16**, 1507 (1977).

31. N. V. Vitanov, B. W. Shore, L. Yatsenko, K. Böhmer, T. Halfmann, T. Rickes, and K. Bergmann. Power broadening revisited: theory and experiment. *Optics Communications* **199**, 117 (2001).

32. M. H. Levitt. *Spin Dynamics: Basics of Nuclear Magnetic Resonance.* Wiley (2008).

33. G. A. Webb. *Nuclear Magnetic Resonance.* Specialist Periodical Reports. Royal Society of Chemistry (2006).

34. P. Callaghan. *Principles of Nuclear Magnetic Resonance Microscopy.* Oxford Science Publications. Oxford University Press (1993).

35. B. Misra and E. C. G. Sudarshan. The Zeno paradox in quantum theory. *Journal of Mathematical Physics* **18**, 756 (1977).

36. Aristotle. Physics. In *Book VI*.

37. H. Nakazato, M. Namiki, S. Pascazio, and H. Rauch. On the quantum Zeno effect. *Physics Letters A* **199**, 27 (1995).

38. R. J. Cook. What are quantum jumps? *Physica Scripta* **T21**, 49 (1988).

39. W. Itano, D. Heinzen, J. Bollinger, and D. Wineland. Quantum Zeno effect. *Physical Review A* **41**, 2295 (1990).

40. A. Venugopalan and R. Ghosh. Decoherence and the quantum Zeno effect. *Physics Letters A* **204**, 11 (1995).

41. H. Nakazato, M. Namiki, S. Pascazio, and H. Rauch. Understanding the quantum Zeno effect. *Physics Letters A* **217**, 203 (1996).

42. G. de Lange, Z. H. Wang, D. Ristè, V. V. Dobrovitski, and R. Hanson. Universal dynamical decoupling of a single solid-state spin from a spin bath. *Science* **330**, 60 (2010).

43. P. Facchi, H. Nakazato, and S. Pascazio. From the quantum Zeno to the inverse quantum Zeno effect. *Physical Review Letters* **86**, 2699 (2001).

44. J. Franson, B. Jacobs, and T. Pittman. Quantum computing using single photons and the Zeno effect. *Physical Review A* **70**, 062302 (2004).

45. L. Zhou, S. Yang, Y.-X. Liu, C. P. Sun, and F. Nori. Quantum Zeno switch for single-photon coherent transport. *Physical Review A* **80**, 062109 (2009).

46. S. Zhang, X.-Q. Shao, L. Chen, Y.-F. Zhao, and K. H. Yeon. Robust $\sqrt{\text{swap}}$-gate on nitrogen-vacancy centres via quantum Zeno dynamics. *Journal of Physics B: Atomic, Molecular and Optical Physics* **44**, 075505 (2011).

47. C. Monroe, D. Meekhof, B. King, S. Jefferts, W. Itano, D. Wineland, and P. Gould. Resolved-Sideband Raman Cooling of a Bound Atom to the 3D Zero-Point Energy. *Physical Review Letters* **75**, 4011 (1995).

48. J. Thomas, P. Hemmer, S. Ezekiel, C. Leiby, R. Picard, and C. Willis. Observation of ramsey fringes using a stimulated, resonance raman transition in a sodium atomic beam. *Physical Review Letters* **48**, 867 (1982).

49. K. Moler, D. Weiss, M. Kasevich, and S. Chu. Theoretical analysis of velocity-selective Raman transitions. *Physical Review A* **45**, 342 (1992).

50. D. J. Wineland, C. Monroe, W. M. Itano, B. E. King, D. Leibfried, D. M. Meekhof, C. Myatt, and C. Wood. Experimental primer on the trapped ion quantum computer. *Fortschritte der Physik* **46**, 363 (1998).

51. D. S. Weiss, B. C. Young, and S. Chu. Precision measurement of $\hbar/m_{Cs}$ based on photon recoil using laser-cooled atoms and atomic interferometry. *Applied Physics B Lasers and Optics* **59**, 217 (1994).

52. E. T. Jaynes and F. W. Cummings. Comparison of quantum and semiclassical radiation theories with application to the beam maser. *Proceedings of the IEEE* **51**, 89 (1963).

53. B. W. Shore and P. L. Knight. The Jaynes-Cummings Model. *Journal of Modern Optics* **40**, 1195 (1993).

54. D. F. Walls and G. J. Milburn. *Quantum Optics*. Springer (2008).

55. E. M. Purcell. Proceedings of the American Physical Society. *Physical Review* **69**, 674 (1946).

56. M. Brune, F. Schmidt-Kaler, A. Maali, J. Dreyer, E. Hagley, J. M. Raimond, and S. Haroche. Quantum Rabi oscillation: a direct test of field quantization in a cavity. *Physical Review Letters* **76**, 1800 (1996).

57. H. Walther, B. T. H. Varcoe, B.-G. Englert, and T. Becker. Cavity quantum electrodynamics. *Reports on Progress in Physics* **69**, 1325 (2006).

58. E. Hagley, X. Maître, G. Nogues, C. Wunderlich, M. Brune, J. M. Raimond, and S. Haroche. Generation of Einstein-Podolsky-Rosen pairs of atoms. *Physical Review Letters* **79**, 1 (1997).

59. P. W. Pinkse, T. Fischer, P. Maunz, and G. Rempe. Trapping an atom with single photons. *Nature* **404**, 365 (2000).

60. P. Maunz, T. Puppe, I. Schuster, N. Syassen, P. W. H. Pinkse, and G. Rempe. Cavity cooling of a single atom. *Nature* **428**, 50 (2004).

61. E. Vetsch, D. Reitz, G. Sagué, R. Schmidt, S. T. Dawkins, and A. Rauschenbeutel. Optical interface created by laser-cooled atoms trapped in the evanescent field surrounding an optical nanofiber. *Physical Review Letters* **104**, 203603 (2010).

62. R. Gehr, J. Volz, G. Dubois, T. Steinmetz, Y. Colombe, B. L. Lev, R. Long, J. Estève, and J. Reichel. Cavity-based single atom preparation and high-

fidelity hyperfine state readout. *Physical Review Letters* **104**, 203602 (2010).

63. M. Grundmann. *The Physics of Semiconductors: An Introduction Including Nanophysics and Applications*. Graduate texts in physics. Springer (2011).

64. Ioffe Institute. New semiconductor materials: characteristics and properties (http://www.ioffe.ru/SVA/NSM/). By time of writing (2013).

65. D. Bimberg, M. Grundmann, and N. N. Ledentsov. *Quantum Dot Heterostructures*. Wiley (1999).

66. U. Woggon. *Optical Properties of Semiconductor Quantum Dots*. No. 136 in Springer Tracts in Modern Physics Series. Springer-Verlag GmbH (1997).

67. D. Leonard, M. Krishnamurthy, C. M. Reaves, S. P. Denbaars, and P. M. Petroff. Direct formation of quantum-sized dots from uniform coherent islands of InGaAs on GaAs surfaces. *Applied Physics Letters* **63**, 3203 (1993).

68. H. Haug and S. W. Koch. *Quantum Theory of the Optical and Electronic Properties of Semiconductors*. World Scientific (2004).

69. J. Gomis-Bresco, S. Dommers, V. V. Temnov, U. Woggon, E. Malic, M. Richter, E. Schöll, and A. Knorr. Impact of coulomb scattering on the ultrafast gain recovery in InGaAs quantum dots. *Physical Review Letters* **101**, 256803 (2008).

70. O. Stier, M. Grundmann, and D. Bimberg. Electronic and optical properties of strained quantum dots modeled by 8-band $k \times p$ theory. *Physical Review B* **59**, 5688 (1999).

71. M.-R. Dachner, E. Malic, M. Richter, A. Carmele, J. Kabuss, A. Wilms, J.-E. Kim, G. Hartmann, J. Wolters, U. Bandelow, and A. Knorr. Theory of carrier and photon dynamics in quantum dot light emitters. *Physica Status Solidi (B)* **247**, 809 (2010).

72. R. J. Warburton, C. Schaflein, D. Haft, F. Bickel, A. Lorke, K. Karrai, J. M. Garcia, W. Schoenfeld, and P. M. Petroff. Optical emission from a charge-tunable quantum ring. *Nature* **405**, 926 (2000).

73. A. P. Alivisatos. Semiconductor clusters, nanocrystals, and quantum dots. *Science* **271**, 933 (1996).

74. N. Gisin and R. Thew. Quantum communication. *Nature Photonics* **1**, 165 (2007).

75. N. D. Mermin. *Quantum Computer Science: An Introduction*. Cambridge University Press (2007).

76. G. Brassard and C. H. Bennett. Quantum cryptography: public key distribution and coin tossing. In *Proc. IEEE International Conference on Computers Systems and Signal Processing* (1984).

77. O. Benson and F. Henneberger. *Semiconductor Quantum Bits*. Pan Stanford Publishing Pte Ltd. (2007).

78. V. Zwiller, T. Aichele, W. Seifert, J. Persson, and O. Benson. Generating visible single photons on demand with single InP quantum dots. *Applied Physics Letters* **82**, 1509 (2003).

79. P. Michler, A. Kiraz, C. Becher, W. V. Schoenfeld, P. M. Petroff, L. Zhang, E. Hu, and A. Imamoglu. A quantum dot single-photon turnstile device. *Science* **290**, 2282 (2000).

80. Z. Yuan, B. E. Kardynal, R. Mark Stevenson, A. J. Shields, C. J. Lobo, K. Cooper, N. S. Beattie, D. A. Ritchie, and M. Pepper. Electrically driven single-photon source. *Science* **295**, 102 (2002).

81. D. Bimberg, E. Stock, A. Lochmann, A. Schliwa, J. A. Tofflinger, W. Unrau, M. Munnix, S. Rodt, V. A. Haisler, A. I. Toropov, A. Bakarov, and A. K. Kalagin. Quantum dots for single- and entangled-photon emitters. *IEEE Photonics Journal* **1**, 58 (2009).

82. A. J. Shields. Semiconductor quantum light sources. *Nature Photonics* **1**, 215 (2007).

83. S. Buckley, K. Rivoire, and J. Vučković. Engineered quantum dot single-photon sources. *Reports on Progress in Physics. Physical Society (Great Britain)* **75**, 126503 (2012).

84. O. Benson, C. Santori, M. Pelton, and Y. Yamamoto. Regulated and entangled photons from a single quantum dot. *Physical Review Letters* **84**, 2513 (2000).

85. N. Akopian, N. Lindner, E. Poem, Y. Berlatzky, J. Avron, D. Gershoni, B. Gerardot, and P. Petroff. Entangled photon pairs from semiconductor quantum dots. *Physical Review Letters* **96**, 130501 (2006).

86. R. M. Stevenson, R. J. Young, P. Atkinson, K. Cooper, D. A. Ritchie, and A. J. Shields. A semiconductor source of triggered entangled photon pairs. *Nature* **439**, 179 (2006).

87. R. Trotta, E. Zallo, C. Ortix, P. Atkinson, J. D. Plumhof, J. van den Brink, A. Rastelli, and O. G. Schmidt. Universal recovery of the energy-level degeneracy of bright excitons in InGaAs quantum dots without a structure symmetry. *Physical Review Letters* **109**, 147401 (2012).

88. D. Loss and D. P. DiVincenzo. Quantum computation with quantum dots. *Physical Review A* **57**, 120 (1998).

89. D. Press, T. D. Ladd, B. Zhang, and Y. Yamamoto. Complete quantum control of a single quantum dot spin using ultrafast optical pulses. *Nature* **456**, 218 (2008).

90. T. D. Ladd, F. Jelezko, R. Laflamme, Y. Nakamura, C. Monroe, and J. L. O'Brien. Quantum computers. *Nature* **464**, 45 (2010).

91. H. Liu, T. Fujisawa, H. Inokawa, Y. Ono, A. Fujiwara, and Y. Hirayama. A gate-defined silicon quantum dot molecule. *Applied Physics Letters* **92**, 222104 (2008).

92. C. B. Simmons, M. Thalakulam, B. M. Rosemeyer, B. J. Van Bael, E. K. Sackmann, D. E. Savage, M. G. Lagally, R. Joynt, M. Friesen, S. N. Coppersmith, and M. A. Eriksson. Charge sensing and controllable tunnel coupling in a Si/SiGe double quantum dot. *Nano Letters* **9**, 3234 (2009).

93. A. Greilich, S. E. Economou, S. Spatzek, D. R. Yakovlev, D. Reuter, A. D. Wieck, T. L. Reinecke, and M. Bayer. Ultrafast optical rotations of electron spins in quantum dots. *Nature Physics* **5**, 262 (2009).

94. F. H. L. Koppens, C. Buizert, K. J. Tielrooij, I. T. Vink, K. C. Nowack, T. Meunier, L. P. Kouwenhoven, and L. M. K. Vandersypen. Driven coherent oscillations of a single electron spin in a quantum dot. *Nature* **442**, 766 (2006).

95. M. Orrit and J. Bernard. Single pentacene molecules detected by fluorescence excitation in a p-terphenyl crystal. *Physical Review Letters* **65**, 2716 (1990).

96. W. E. Moerner and D. P. Fromm. Methods of single-molecule fluorescence spectroscopy and microscopy. *Review of Scientific Instruments* **74**, 3597 (2003).

97. C. Toninelli, K. Early, J. Bremi, A. Renn, S. Götzinger, and V. Sandoghdar. Near-infrared single-photons from aligned molecules in ultrathin crystalline films at room temperature. *Optics Express* **18**, 6579 (2010).

98. W. E. Moerner. Single-photon sources based on single molecules in solids. *New Journal of Physics* **6**, 88 (2004).

99. C. Brunel, B. Lounis, P. Tamarat, and M. Orrit. Triggered source of single photons based on controlled single molecule fluorescence. *Physical Review Letters* **83**, 2722 (1999).

100. B. Lounis and W. E. Moerner. Single photons on demand from a single molecule at room temperature. *Nature* **407**, 491 (2000).

101. Y. Tian, P. Navarro, B. Kozankiewicz, and M. Orrit. Spectral diffusion of single dibenzoterrylene molecules in 2,3-dimethylanthracene. *Chemphyschem: A European Journal of Chemical Physics and Physical Chemistry* **13**, 3510 (2012).

102. A. A. L. Nicolet, C. Hofmann, M. A. Kol'chenko, B. Kozankiewicz, and M. Orrit. Single dibenzoterrylene molecules in an anthracene crystal: spectroscopy and photophysics. *Chemphyschem: A European Journal of Chemical Physics and Physical Chemistry* **8**, 1215 (2007).

103. F. Jelezko, P. H. Tamarat, B. Lounis, and M. Orrit. Dibenzoterrylene in naphthalene: a new crystalline system for single molecule spectroscopy in the near infrared. *The Journal of Physical Chemistry* **100**, 13892 (1996).

104. H. Bock, K. Gharagozloo-Hubmann, M. Sievert, T. Prisner, and Z. Havlas. Single crystals of an ionic anthracene aggregate with a triplet ground state. *Nature* **404**, 267 (2000).

105. Th. Basché, S. Kummer, and C. Bräuchle. Direct spectroscopic observation of quantum jumps of a single molecule. *Nature* **373**, 132 (1995).

106. J. Wrachtrup, C. von Borczyskowski, J. Bernard, M. Orritt, and R. Brown. Optical detection of magnetic resonance in a single molecule. *Nature* **363**, 244 (1993).

107. J. Köhler, J. A. J. M. Disselhorst, M. C. J. M. Donckers, E. J. J. Groenen, J. Schmidt, and W. E. Moerner. Magnetic resonance of a single molecular spin. *Nature* **363**, 242 (1993).

108. G. Feher. Observation of nuclear magnetic resonances via the electron spin resonance line. *Physical Review* **103**, 834 (1956).

109. I. Aharonovich, S. Castelletto, D. A. Simpson, C.-H. Su, A. D. Greentree, and S. Prawer. Diamond-based single-photon emitters. *Reports on Progress in Physics* **74**, 076501 (2011).

110. A. Beveratos, R. Brouri, T. Gacoin, J.-P. Poizat, and P. Grangier. Nonclassical radiation from diamond nanocrystals. *Physical Review A* **64**, 061802 (2001).

111. E. Neu, D. Steinmetz, J. Riedrich-Möller, S. Gsell, M. Fischer, M. Schreck, and C. Becher. Single photon emission from silicon-vacancy colour centres in chemical vapour deposition nano-diamonds on iridium. *New Journal of Physics* **13**, 025012 (2011).

112. C. Wang, C. Kurtsiefer, H. Weinfurter, and B. Burchard. Single photon emission from SiV centres in diamond produced by ion implantation. *Journal of Physics B: Atomic, Molecular and Optical Physics* **39**, 37 (2006).

113. E. Wu, J. R. Rabeau, G. Roger, F. Treussart, H. Zeng, P. Grangier, S. Prawer, and J.-F. Roch. Room temperature triggered single-photon source in the near infrared. *New Journal of Physics* **9**, 434 (2007).

114. T. Gaebel, I. Popa, A. Gruber, M. Domhan, F. Jelezko, and J. Wrachtrup. Stable single-photon source in the near infrared. *New Journal of Physics* **6**, 98 (2004).

115. R. Hanson and D. D. Awschalom. Coherent manipulation of single spins in semiconductors. *Nature* **453**, 1043 (2008).

116. J. L. O'Brien, A. Furusawa, and J. Vučković. Photonic quantum technologies. *Nature Photonics* **3**, 687 (2009).

117. I. Aharonovich, A. D. Greentree, and S. Prawer. Diamond photonics. *Nature Photonics* **5**, 397 (2011).

118. J. Wrachtrup and F. Jelezko. Processing quantum information in diamond. *Journal of Physics: Condensed Matter* **18**, S807 (2006).

119. F. Shi, X. Rong, N. Xu, Y. Wang, J. Wu, B. Chong, X. Peng, J. Kniepert, R.-S. Schoenfeld, W. Harneit, M. Feng, and J. Du. Room-temperature implementation of the Deutsch-Jozsa algorithm with a single Electronic spin in diamond. *Physical Review Letters* **105**, 040504 (2010).

120. P. Neumann, R. Kolesov, B. Naydenov, J. Beck, F. Rempp, M. Steiner, V. Jacques, G. Balasubramanian, M. L. Markham, D. J. Twitchen, S. Pezzagna, J. Meijer, J. Twamley, F. Jelezko, and J. Wrachtrup. Quantum register based on coupled electron spins in a room-temperature solid. *Nature Physics* **6**, 249 (2010).

121. E. Togan, Y. Chu, A. S. Trifonov, L. Jiang, J. Maze, L. Childress, M. V. G. Dutt, A. S. Sø rensen, P. R. Hemmer, A. S. Zibrov, and M. D. Lukin. Quantum entanglement between an optical photon and a solid-state spin qubit. *Nature* **466**, 730 (2010).

122. F. Dolde, I. Jakobi, B. Naydenov, N. Zhao, S. Pezzagna, C. Trautmann, J. Meijer, P. Neumann, F. Jelezko, and J. Wrachtrup. Room-temperature entanglement between single defect spins in diamond. *Nature Physics* **9**, 13 (2013).

123. H. Bernien, B. Hensen, W. Pfaff, G. Koolstra, M. S. Blok, L. Robledo, T. H. Taminiau, M. Markham, D. J. Twitchen, L. Childress, and R. Hanson. Heralded entanglement between solid-state qubits separated by three metres. *Nature* **497**, 86 (2013).

124. C. Santori, P. Tamarat, P. Neumann, J. Wrachtrup, D. Fattal, R. Beausoleil, J. Rabeau, P. Olivero, A. Greentree, S. Prawer, F. Jelezko, and P. Hemmer.

Coherent population trapping of single spins in diamond under optical excitation. *Physical Review Letters* **97**, 247401 (2006).

125. F. Hilser and G. Burkard. All-optical control of the spin state in the NV⁻ center in diamond. *Physical Review B* **86**, 125204 (2012).

126. C. G. Yale, B. B. Buckley, D. J. Christle, G. B. Burkard, F. J. Heremans, L. C. Bassett, and D. D. Awschalom. All-optical control of a solid-state spin using coherent dark states. *Proceedings of the National Academy of Sciences of the United States of America* **110**, 7595 (2013).

127. V. M. Acosta, K. Jensen, C. Santori, D. Budker, and R. G. Beausoleil. Electromagnetically induced transparency in a diamond spin ensemble enables all-optical electromagnetic field sensing. *Physical Review Letters* **110**, 213605 (2013).

128. J. Wolters, A. W. Schell, G. Kewes, N. Nüsse, M. Schoengen, H. Döscher, T. Hannappel, B. Löchel, M. Barth, and O. Benson. Enhancement of the zero phonon line emission from a single nitrogen vacancy center in a nanodiamond via coupling to a photonic crystal cavity. *Applied Physics Letters* **97**, 141108 (2010).

129. J. Wolters, G. Kewes, A. W. Schell, N. Nüsse, M. Schoengen, B. Löchel, T. Hanke, R. Bratschitsch, A. Leitenstorfer, T. Aichele, and O. Benson. Coupling of single nitrogen-vacancy defect centers in diamond nanocrystals to optical antennas and photonic crystal cavities. *Physica Status Solidi (B)* **249**, 918 (2012).

130. T. van der Sar, J. Hagemeier, W. Pfaff, E. C. Heeres, S. M. Thon, H. Kim, P. M. Petroff, T. H. Oosterkamp, D. Bouwmeester, and R. Hanson. Deterministic nanoassembly of a coupled quantum emitter-photonic crystal cavity system. *Applied Physics Letters* **98**, 193103 (2011).

131. O. Benson. Assembly of hybrid photonic architectures from nanophotonic constituents. *Nature* **480**, 193 (2011).

132. D. Englund, B. Shields, K. Rivoire, F. Hatami, J. Vučković, H. Park, and M. D. Lukin. Deterministic coupling of a single nitrogen vacancy center to a photonic crystal cavity. *Nano Letters* **10**, 3922 (2010).

133. J.-P. Boudou, P. A. Curmi, F. Jelezko, J. Wrachtrup, P. Aubert, M. Sennour, G. Balasubramanian, R. Reuter, A. Thorel, and E. Gaffet. High yield fabrication of fluorescent nanodiamonds. *Nanotechnology* **20**, 235602 (2009).

134. A. Pentecost, S. Gour, V. Mochalin, I. Knoke, and Y. Gogotsi. Deaggregation of nanodiamond powders using salt- and sugar-assisted milling. *ACS Applied Materials & Interfaces* **2**, 3289 (2010).

135. A. Krueger. Beyond the shine: recent progress in applications of nanodiamond. *Journal of Materials Chemistry* **21**, 12571 (2011).

136. E. Neu, C. Hepp, M. Hauschild, S. Gsell, M. Fischer, H. Sternschulte, D. Steinmüller-Nethl, M. Schreck, and C. Becher. Low-temperature investigations of single silicon vacancy colour centres in diamond. *New Journal of Physics* **15**, 043005 (2013).

137. E. Neu, M. Agio, and C. Becher. Photophysics of single silicon vacancy centers in diamond: implications for single photon emission. *Optics Express* **20**, 19956 (2012).

138. I. Aharonovich, S. Castelletto, D. A. Simpson, A. Stacey, J. McCallum, A. D. Greentree, and S. Prawer. Two-level ultrabright single photon emission from diamond nanocrystals. *Nano Letters* **9**, 3191 (2009).

139. I. Aharonovich, S. Castelletto, B. C. Johnson, J. C. McCallum, D. A. Simpson, A. D. Greentree, and S. Prawer. Chromium single-photon emitters in diamond fabricated by ion implantation. *Physical Review B* **81**, 121201 (2010).

140. I. Aharonovich, S. Castelletto, D. A. Simpson, A. D. Greentree, and S. Prawer. Photophysics of chromium-related diamond single-photon emitters. *Physical Review A* **81**, 043813 (2010).

141. I. Aharonovich, S. Castelletto, B. C. Johnson, J. C. McCallum, and S. Prawer. Engineering chromium-related single photon emitters in single crystal diamonds. *New Journal of Physics* **13**, 045015 (2011).

142. P. Siyushev, V. Jacques, I. Aharonovich, F. Kaiser, T. Müller, L. Lombez, M. Atatüre, S. Castelletto, S. Prawer, F. Jelezko, and J. Wrachtrup. Low-temperature optical characterization of a near-infrared single-photon emitter in nanodiamonds. *New Journal of Physics* **11**, 113029 (2009).

143. A. Edmonds, M. Newton, P. Martineau, D. Twitchen, and S. Williams. Electron paramagnetic resonance studies of silicon-related defects in diamond. *Physical Review B* **77**, 245205 (2008).

144. G. Sittas, H. Kanda, I. Kiflawi, and P. M. Spear. Growth and characterization of Si-doped diamond single crystals grown by the HTHP method. *Diamond and Related Materials* **5**, 866 (1996).

145. C. Clark, H. Kanda, I. Kiflawi, and G. Sittas. Silicon defects in diamond. *Physical Review B* **51**, 16681 (1995).

146. K.-M. C. Fu, C. Santori, P. E. Barclay, and R. G. Beausoleil. Conversion of neutral nitrogen-vacancy centers to negatively charged nitrogen-vacancy centers through selective oxidation. *Applied Physics Letters* **96**, 121907 (2010).

147. L. Rondin, G. Dantelle, A. Slablab, F. Grosshans, F. Treussart, P. Bergonzo, S. Perruchas, T. Gacoin, M. Chaigneau, H.-C. Chang, V. Jacques, and J.-F. Roch. Surface-induced charge state conversion of nitrogen-vacancy defects in nanodiamonds. *Physical Review B* **82**, 115449 (2010).

148. P. Siyushev, H. Pinto, M. Vörös, A. Gali, F. Jelezko, and J. Wrachtrup. Optically controlled switching of the charge state of a single nitrogen-vacancy center in diamond at cryogenic temperatures. *Physical Review Letters* **110**, 167402 (2013).

149. S. M. Blinder. Lecture Notes on Quantum Chemistry, Chapter 12 (Molecular Symmetry). University of Michigan (2002).

150. L. Robledo, L. Childress, H. Bernien, B. Hensen, P. F. A. Alkemade, and R. Hanson. High-fidelity projective read-out of a solid-state spin quantum register. *Nature* **477**, 574 (2011).

151. A. Sipahigil, M. L. Goldman, E. Togan, Y. Chu, M. Markham, D. J. Twitchen, A. S. Zibrov, A. Kubanek, and M. D. Lukin. Quantum interference of single photons from remote nitrogen-vacancy centers in diamond. *Physical Review Letters* **108**, 143601 (2012).

152. H. Bernien, L. Childress, L. Robledo, M. Markham, D. Twitchen, and R. Hanson. Two-photon quantum interference from separate nitrogen vacancy centers in diamond. *Physical Review Letters* **108**, 043604 (2012).

153. Y. Shen, T. Sweeney, and H. Wang. Zero-phonon linewidth of single nitrogen vacancy centers in diamond nanocrystals. *Physical Review B* **77**, 033201 (2008).

154. J. Wolters, N. Sadzak, A. W. Schell, T. Schröder, and O. Benson. Measurement of the ultrafast spectral diffusion of the optical transition of nitrogen vacancy centers in nano-size diamond using correlation interferometry. *Physical Review Letters* **110**, 027401 (2013).

155. K.-M. C. Fu, C. Santori, P. E. Barclay, and R. G. Beausoleil. Conversion of neutral nitrogen-vacancy centers to negatively charged nitrogen-vacancy centers through selective oxidation. *Applied Physics Letters* **96**, 121907 (2010).

156. A. Majumdar, E. D. Kim, and J. Vučković. Effect of photogenerated carriers on the spectral diffusion of a quantum dot coupled to a photonic crystal cavity. *Physical Review B* **84**, 195304 (2011).

157. C. Zander, J. Enderlein, and R. A. Keller. *Single Molecule Detection in Solution: Methods and Applications*. Wiley-VCH (2002).

158. R. G. Farrer. On the substitutional nitrogen donor in diamond. *Solid State Communications* **7**, 685 (1969).

159. J. Rosa, M. Vaněček, M. Nesládek, and L. M. Stals. Photoionization cross-section of dominant defects in CVD diamond. *Diamond and Related Materials* **8**, 175 (1999).

160. V. M. Acosta, C. Santori, A. Faraon, Z. Huang, K.-M. C. Fu, A. Stacey, D. A. Simpson, K. Ganesan, S. Tomljenovic-Hanic, A. D. Greentree, S. Prawer, and R. G. Beausoleil. Dynamic stabilization of the optical resonances of single nitrogen-vacancy centers in diamond. *Physical Review Letters* **108**, 206401 (2012).

161. Ph. Tamarat, T. Gaebel, J. Rabeau, M. Khan, A. Greentree, H. Wilson, L. Hollenberg, S. Prawer, P. Hemmer, F. Jelezko, and J. Wrachtrup. Stark shift control of single optical centers in diamond. *Physical Review Letters* **97**, 083002 (2006).

162. D. Bimberg. *Semiconductor Nanostructures*. NanoScience and Technology. Springer-Verlag (2008).

163. G. Sallen, A. Tribu, T. Aichele, R. André, L. Besombes, C. Bougerol, M. Richard, S. Tatarenko, K. Kheng, and J.-Ph. Poizat. Subnanosecond spectral diffusion measurement using photon correlation. *Nature Photonics* **4**, 696 (2010).

164. X. Brokmann, M. Bawendi, L. Coolen, and J.-P. Hermier. Photon-correlation Fourier spectroscopy. *Optics Express* **14**, 6333 (2006).

165. L. Coolen, X. Brokmann, P. Spinicelli, and J.-P. Hermier. Emission characterization of a single CdSe-ZnS nanocrystal with high temporal and spectral resolution by photon-correlation Fourier spectroscopy. *Physical Review Letters* **100**, 027403 (2008).

166. M. Orrit. Single-photon sources: frequency jitter of a nano-emitter. *Nature Photonics* **4**, 667 (2010).

167. T. Schröder, F. Gädeke, M. J. Banholzer, and O. Benson. Ultrabright and efficient single-photon generation based on nitrogen-vacancy centres in nanodiamonds on a solid immersion lens. *New Journal of Physics* **13**, 055017 (2011).

168. G. Waldherr, J. Beck, M. Steiner, P. Neumann, A. Gali, Th. Frauenheim, F. Jelezko, and J. Wrachtrup. Dark states of single nitrogen-vacancy centers in diamond unraveled by single shot NMR. *Physical Review Letters* **106**, 157601 (2011).

169. H. Robinson and B. Goldberg. Light-induced spectral diffusion in single self-assembled quantum dots. *Physical Review B* **61**, R5086 (2000).

170. E. Knill, R. Laflamme, and G. J. Milburn. A scheme for efficient quantum computation with linear optics. *Nature* **409**, 46 (2001).

171. N. Manson, J. Harrison, and M. Sellars. Nitrogen-vacancy center in diamond: Model of the electronic structure and associated dynamics. *Physical Review B* **74**, 104303 (2006).

172. A. Batalov, V. Jacques, F. Kaiser, P. Siyushev, P. Neumann, L. Rogers, R. McMurtrie, N. Manson, F. Jelezko, and J. Wrachtrup. Low temperature studies of the excited-state structure of negatively charged nitrogen-vacancy color centers in diamond. *Physical Review Letters* **102**, 195506 (2009).

173. A. Lenef and S. C. Rand. Electronic structure of the N-V center in diamond: theory. *Physical review. B, Condensed matter* **53**, 13441 (1996).

174. G. Fuchs, V. Dobrovitski, R. Hanson, A. Batra, C. Weis, T. Schenkel, and D. Awschalom. Excited-state spectroscopy using single spin manipulation in diamond. *Physical Review Letters* **101**, 117601 (2008).

175. J. R. Maze, A. Gali, E. Togan, Y. Chu, A. Trifonov, E. Kaxiras, and M. D. Lukin. Properties of nitrogen-vacancy centers in diamond: the group theoretic approach. *New Journal of Physics* **13**, 025025 (2011).

176. L. T. Hall, G. C. G. Beart, E. A. Thomas, D. A. Simpson, L. P. McGuinness, J. H. Cole, J. H. Manton, R. E. Scholten, F. Jelezko, J. Wrachtrup, S. Petrou, and L. C. L. Hollenberg. High spatial and temporal resolution wide-field imaging of neuron activity using quantum NV-diamond. *Scientific Reports* **2**, 401 (2012).

177. K. Fang, V. M. Acosta, C. Santori, Z. Huang, K. M. Itoh, H. Watanabe, S. Shikata, and R. G. Beausoleil. High-sensitivity magnetometry based on quantum beats in diamond nitrogen-vacancy centers. *Physical Review Letters* **110**, 130802 (2013).

178. R. S. Schönfeld. *Optical readout of single spins for quantum computing and magnetic sensing.* PhD thesis, Freie Universität Berlin (2011).

179. A. Hegyi and E. Yablonovitch. Molecular imaging by optically detected electron spin resonance of nitrogen-vacancies in nanodiamonds. *Nano Letters* **13**, 1173 (2013).

180. N. D. P. Lai, D. Zheng, F. Jelezko, F. Treussart, and J.-F. Roch. Influence of a static magnetic field on the photoluminescence of an ensemble of nitrogen-vacancy color centers in a diamond single-crystal. *Applied Physics Letters*, **95**, 133101 (2009).

181. J. Maze Rios. *Quantum manipulation of nitrogen-vacancy centers in diamond: from basic properties to applications.* PhD thesis, Harvard University (2010).

182. S.-Y. Lee, M. Widmann, T. Rendler, M. W. Doherty, T. M. Babinec, S. Yang, M. Eyer, P. Siyushev, B. J. M. Hausmann, M. Loncar, Z. Bodrog, A. Gali, N. B. Manson, H. Fedder, and J. Wrachtrup. Readout and control of a single nuclear spin with a metastable electron spin ancilla. *Nature Nanotechnology* **8**, 487 (2013).

183. A. Dréau, P. Spinicelli, J. R. Maze, J.-F. Roch, and V. Jacques. Single-shot readout of multiple nuclear spin qubits in diamond under ambient conditions. *Physical Review Letters* **110**, 060502 (2013).

184. L. M. Vandersypen, M. Steffen, G. Breyta, C. S. Yannoni, M. H. Sherwood, and I. L. Chuang. Experimental realization of Shor's quantum factoring algorithm using nuclear magnetic resonance. *Nature* **414**, 883 (2001).

185. F. Jelezko, T. Gaebel, I. Popa, M. Domhan, A. Gruber, and J. Wrachtrup. Observation of coherent oscillation of a single nuclear spin and realization of a two-qubit conditional quantum gate. *Physical Review Letters* **93**, 130501 (2004).

186. M. V. G. Dutt, L. Childress, L. Jiang, E. Togan, J. Maze, F. Jelezko, A. S. Zibrov, P. R. Hemmer, and M. D. Lukin. Quantum register based on individual electronic and nuclear spin qubits in diamond. *Science* **316**, 1312 (2007).

187. A. Laraoui, J. S. Hodges, C. A. Ryan, and C. A. Meriles. Diamond nitrogen-vacancy center as a probe of random fluctuations in a nuclear spin ensemble. *Physical Review B* **84**, 104301 (2011).

188. A. Bermudez, F. Jelezko, M. B. Plenio, and A. Retzker. Electron-mediated nuclear-spin interactions between distant nitrogen-vacancy centers. *Physical Review Letters* **107**, 150503 (2011).

189. E. Samuel Reich. Quantum paradox seen in diamond. *Nature*, (2013).

190. J. Wolters, M. Strauß, R. S. Schoenfeld, and O. Benson. Quantum Zeno phenomenon on a single solid-state spin. *Physical Review A* **88**, 020101 (2013).

191. L. Robledo, H. Bernien, T. van der Sar, and R. Hanson. Spin dynamics in the optical cycle of single nitrogen-vacancy centres in diamond. *New Journal of Physics* **13**, 025013 (2011).

192. Ph. Tamarat, N. B. Manson, J. P. Harrison, R. L. McMurtrie, A. Nizovtsev, C. Santori, R. G. Beausoleil, P. Neumann, T. Gaebel, F. Jelezko, P. Hemmer, and J. Wrachtrup. Spin-flip and spin-conserving optical transitions of

the nitrogen-vacancy centre in diamond. *New Journal of Physics* **10**, 045004 (2008).

193. G. D. Fuchs, V. V. Dobrovitski, D. M. Toyli, F. J. Heremans, C. D. Weis, T. Schenkel, and D. D. Awschalom. Excited-state spin coherence of a single nitrogen-vacancy centre in diamond. *Nature Physics* **6**, 668 (2010).

194. L. J. Rogers, R. L. McMurtrie, M. J. Sellars, and N. B. Manson. Time-averaging within the excited state of the nitrogen-vacancy centre in diamond. *New Journal of Physics* **11**, 063007 (2009).

195. G. L. González and M. N. Leuenberger. The dynamics of the optically driven Lambda transition of the ^{15}N-V-center in diamond. *Nanotechnology* **21**, 274020 (2010).

196. P. C. Maurer, J. R. Maze, P. L. Stanwix, L. Jiang, A. V. Gorshkov, A. A. Zibrov, B. Harke, J. S. Hodges, A. S. Zibrov, A. Yacoby, D. Twitchen, S. W. Hell, R. L. Walsworth, and M. D. Lukin. Far-field optical imaging and manipulation of individual spins with nanoscale resolution. *Nature Physics* **6**, 912 (2010).

197. P. Maletinsky, S. Hong, M. S. Grinolds, B. Hausmann, M. D. Lukin, R. L. Walsworth, M. Loncar, and A. Yacoby. A robust scanning diamond sensor for nanoscale imaging with single nitrogen-vacancy centres. *Nature Nanotechnology* **7**, 320 (2012).

198. A. W. Schell, J. Kaschke, J. Fischer, R. Henze, J. Wolters, M. Wegener, and O. Benson. Three-dimensional quantum photonic elements based on single nitrogen vacancy-centres in laser-written microstructures. *Scientific Reports* **3**, 1577 (2013).

199. J. D. Joannopoulos, S. G. Johnson, J. N. Winn, and R. D. Meade. *Molding the flow of light*. Princeton University Press (2007).

200. F. Bloch. Über die Quantenmechanik der Elektronen in Kristallgittern. *Zeitschrift fuer Physik* **52**, 555 (1929).

201. C. Kittel. *Introduction to Solid State Physics*. Wiley (2004).

202. A. R. Franklin and R. Newman. Shaped electroluminescent GaAs diodes. *Journal of Applied Physics* **35**, 1153 (1964).

203. S. V. Galginaitis. Improving the external efficiency of electroluminescent diodes. *Journal of Applied Physics* **36**, 460 (1965).

204. L. Novotny and B. Hecht. *Principles of Nano-Optics*. Cambridge University Press (2012).

205. P. Aigrain. Light emission from injecting contacts on germanium in the 2μ to 6μ band. *Physica* **20**, 1010 (1954).

206. S. M. Mansfield and G. S. Kino. Solid immersion microscope. *Applied Physics Letters* **57**, 2615 (1990).

207. Q. Wu, G. D. Feke, R. D. Grober, and L. P. Ghislain. Realization of numerical aperture 2.0 using a gallium phosphide solid immersion lens. *Applied Physics Letters* **75**, 4064 (1999).

208. P. Siyushev, F. Kaiser, V. Jacques, I. Gerhardt, S. Bischof, H. Fedder, J. Dodson, M. Markham, D. Twitchen, F. Jelezko, and J. Wrachtrup. Monolithic diamond optics for single photon detection. *Applied Physics Letters* **97**, 241902 (2010).

209. T. Schröder, F. Gädeke, M. J. Banholzer, and O. Benson. Ultrabright and efficient single-photon generation based on nitrogen-vacancy centres in nanodiamonds on a solid immersion lens. *New Journal of Physics* **13**, 055017 (2011).

210. K. J. Ebeling. *Integrated Optoelectronics: Waveguide Optics, Photonics, Semiconductors*. Springer-Verlag (1993).

211. S. Johnson and J. Joannopoulos. Block-iterative frequency-domain methods for Maxwell's equations in a planewave basis. *Optics Express* **8**, 173 (2001).

212. C. Xiong, W. H. P. Pernice, M. Li, and H. X. Tang. High performance nanophotonic circuits based on partially buried horizontal slot waveguides. *Optics Express* **18**, 20690 (2010).

213. P. Rath, N. Gruhler, S. Khasminskaya, C. Nebel, C. Wild, and W. H. P. Pernice. Waferscale nanophotonic circuits made from diamond-on-insulator substrates. *Optics Express* **21**, 11031 (2013).

214. J. Hecht. *Understanding Fiber Optics*. Prentice Hall, 2006.

215. C. Wuttke, M. Becker, S. Brückner, M. Rothhardt, and A. Rauschenbeutel. Nanofiber Fabry-Perot microresonator for non-linear optics and cavity quantum electrodynamics. *Optics Letters* **37**, 1949 (2012).

216. T. Schröder, M. Fujiwara, T. Noda, H.-Q. Zhao, O. Benson, and S. Takeuchi. A nanodiamond-tapered fiber system with high single-mode coupling efficiency. *Optics Express* **20**, 10490 (2012).

217. S. V. Gaponenko. *Introduction to Nanophotonics*. Cambridge University Press (2010).

218. D. K. Armani, T. J. Kippenberg, S. M. Spillane, and K. J. Vahala. Ultra-high-Q toroid microcavity on a chip. *Nature* **421**, 925 (2003).

219. K. J. Vahala. Optical microcavities. *Nature* **424**, 839 (2003).

220. R. Henze, J. M. Ward, and O. Benson. Temperature independent tuning of whispering gallery modes in a cryogenic environment. *Optics Express* **21**, 675 (2013).

221. A. V. Kavokin, J. J. Baumberg, G. Malpuech, and F. P. Laussy. *Microcavities*. Oxford Science Publications (2007).

222. T. Kippenberg, A. Tchebotareva, J. Kalkman, A. Polman, and K. Vahala. Purcell-factor-enhanced scattering from Si nanocrystals in an optical microcavity. *Physical Review Letters* **103**(2), 027406 (2009).

223. M. Deubel, G. von Freymann, M. Wegener, S. Pereira, K. Busch, and C. M. Soukoulis. Direct laser writing of three-dimensional photonic-crystal templates for telecommunications. *Nature Materials* **3**, 444 (2004).

224. S. Kawata, H. B. Sun, T. Tanaka, and K. Takada. Finer features for functional microdevices. *Nature* **412**, 697 (2001).

225. Z.o-P. Liu, Y. Li, Y.-F. Xiao, B.-B. Li, X.-F. Jiang, Y. Qin, X.-B. Feng, H. Yang, and Q. Gong. Direct laser writing of whispering gallery microcavities by two-photon polymerization. *Applied Physics Letters* **97**, 211105 (2010).

226. T. Grossmann, S. Schleede, M. Hauser, T. Beck, M. Thiel, G. von Freymann, T. Mappes, and H. Kalt. Direct laser writing for active and passive high-Q polymer microdisks on silicon. *Optics Express* **19**, 11451 (2011).

227. C.-W. Lee, S. Pagliara, U. Keyser, and J. J. Baumberg. Perpendicular coupling to in-plane photonics using arc waveguides fabricated via two-photon polymerization. *Applied Physics Letters* **100**, 171102 (2012).

228. E. Yablonovitch, T. Gmitter, and K. Leung. Photonic band structure: the face-centered-cubic case employing nonspherical atoms. *Physical Review Letters* **67**, 2295 (1991).

229. H. Sözüer, J. Haus, and R. Inguva. Photonic bands: convergence problems with the plane-wave method. *Physical Review B* **45**, 13962 (1992).

230. K. Ho, C. Chan, and C. Soukoulis. Existence of a photonic gap in periodic dielectric structures. *Physical Review Letters*, **65**(25), 3152–3155 (1990).

231. B. Gralak, G. Tayeb, and S. Enoch. Morpho butterflies wings color modeled with lamellar grating theory. *Optics Express* **9**, 567 (2001).

232. J. Zi, X. Yu, Y. Li, X. Hu, C. Xu, X. Wang, X. Liu, and R. Fu. Coloration strategies in peacock feathers. *Proceedings of the National Academy of Sciences of the United States of America* **100**, 12576 (2003).

233. S. Y. Lin, J. G. Fleming, D. L. Hetherington, B. K. Smith, W. Zubrzycki, S. R. Kurtz, and J. Bur. A three-dimensional photonic crystal operating at infrared wavelengths. *Nature* **394**, 251 (1998).

234. M. R. Jorgensen, J. W. Galusha, and M. H. Bartl. Strongly modified spontaneous emission rates in diamond-structured photonic crystals. *Physical Review Letters* **107**, 143902 (2011).

235. A. Tandaechanurat, S. Ishida, D. Guimard, M. Nomura, S. Iwamoto, and Y. Arakawa. Lasing oscillation in a three-dimensional photonic crystal nanocavity with a complete bandgap. *Nature Photonics* **5**, 175 (2010).

236. S. Johnson, S. Fan, P. Villeneuve, J. Joannopoulos, and L. Kolodziejski. Guided modes in photonic crystal slabs. *Physical Review B* **60**, 5751 (1999).

237. M. Modreanu, N. Tomozeiu, P. Cosmin, and M. Gartner. Optical properties of LPCVD silicon oxynitride. *Thin Solid Films* **337**, 82 (1999).

238. O. Madelung, U. Rössler, and M. Schulz, editors. *Group IV Elements, IV-IV and III-V Compounds. Part b: Electronic, Transport, Optical and Other Properties*, Vol. b of *Landolt-Börnstein – Group III Condensed Matter.* Springer-Verlag (2002).

239. G. Samara. Temperature and pressure dependences of the dielectric constants of semiconductors. *Physical Review B* **27**, 3494 (1983).

240. O. Madelung, U. Rössler, and M. Schulz, editors. *Group IV Elements, IV-IV and III-V Compounds. Part b: Electronic, Transport, Optical and Other Properties*, Vol. 41A1b of *Landolt-Börnstein – Group III Condensed Matter.* Springer-Verlag, Berlin/Heidelberg, 2002.

241. Z. Cui. *Micro-Nanofabrication: Technologies and Applications.* Springer (2010).

242. H. Döscher, T. Hannappel, B. Kunert, A.s Beyer, K. Volz, and W. Stolz. In situ verification of single-domain III-V on Si(100) growth via metal-organic vapor phase epitaxy. *Applied Physics Letters* **93**, 172110 (2008).

243. H. Döscher and T. Hannappel. In situ reflection anisotropy spectroscopy analysis of heteroepitaxial GaP films grown on Si(100). *Journal of Applied Physics* **107**, 123523 (2010).

244. Chr. Balzer, Th. Hannemann, D. Reiß, Chr. Wunderlich, W. Neuhauser, and P. E. Toschek. A relaxationless demonstration of the Quantum Zeno paradox on an individual atom. *Optics Communications* **211**, 175 (2002).

245. S. Johnson, P. Villeneuve, S. Fan, and J. Joannopoulos. Linear waveguides in photonic-crystal slabs. *Physical Review B* **62**, 8212 (2000).

246. A. Mekis, J. C. Chen, I. Kurland, S. Fan, P. R. Villeneuve, and J. D. Joannopoulos. High transmission through sharp bends in photonic crystal waveguides. *Physical Review Letters* **77**, 3787 (1996).

247. A. Chutinan and S. Noda. Waveguides and waveguide bends in two-dimensional photonic crystal slabs. *Physical Review B* **62**, 4488 (2000).

248. S. Lin. Experimental demonstration of guiding and bending of electromagnetic waves in a photonic crystal. *Science* **282**, 274 (1998).

249. L. Frandsen, A. Harpøth, P. Borel, M. Kristensen, J. Jensen, and O. Sigmund. Broadband photonic crystal waveguide 60 degrees bend obtained utilizing topology optimization. *Optics Express* **12**, 5916 (2004).

250. A. Mekis, S. Fan, and J. Joannopoulos. Bound states in photonic crystal waveguides and waveguide bends. *Physical Review B* **58**, 4809 (1998).

251. M. Notomi, A. Shinya, K. Yamada, J.-I. Takahashi, C. Takahashi, and I. Yokohama. Structural tuning of guiding modes of line-defect waveguides of silicon-on-insulator photonic crystal slabs. *IEEE Journal of Quantum Electronics* **38**, 736 (2002).

252. J. Wolters, N. Nikolay, M. Schoengen, A. W. Schell, J. Probst, B. Löchel, and O. Benson. Thermo-optical response of photonic crystal cavities operating in the visible spectral range. *Nanotechnology* **24**, 315204 (2013).

253. K. S. Yee. Numerical solution of initial boundary value problems involving maxwell's equations in isotropic media. *IEEE Transactions on Antennas and Propagation* **14**, 302 (1966).

254. A. F. Oskooi, D. Roundy, M. Ibanescu, P. Bermel, J. D. Joannopoulos, and S. G. Johnson. Meep: a flexible free-software package for electromagnetic simulations by the FDTD method. *Computer Physics Communications* **181**, 687 (2010).

255. A. R. A. Chalcraft, S. Lam, D. O'Brien, T. F. Krauss, M. Sahin, D. Szymanski, D. Sanvitto, R. Oulton, M. S. Skolnick, A. M. Fox, D. M. Whittaker, H.-Y. Liu, and M. Hopkinson. Mode structure of the L3 photonic crystal cavity. *Applied Physics Letters* **90**, 241117 (2007).

256. V. A. Mandelshtam and H. S. Taylor. Harmonic inversion of time signals and its applications. *The Journal of Chemical Physics* **107**, 6756 (1997).

257. J. Vuckovic, M. Loncar, H. Mabuchi, and A. Scherer. Optimization of the Q factor in photonic crystal microcavities. *IEEE Journal of Quantum Electronics* **38**, 850 (2002).

258. M. Barth. *Hybrid Nanophotonic Elements and Sensing Devices based on Photonic Crystal Structures*. PhD thesis, Humboldt-Universität zu Berlin (2010).

259. P. Lalanne, C. Sauvan, and J. P. Hugonin. Photon confinement in photonic crystal nanocavities. *Laser & Photonics Review* **2**, 514 (2008).

260. Y. Akahane, T. Asano, B.-S. Song, and S. Noda. High-Q photonic nanocavity in a two-dimensional photonic crystal. *Nature* **425**, 944 (2003).

261. M. Barth, J. Kouba, J. Stingl, B. Löchel, and O. Benson. Modification of visible spontaneous emission with silicon nitride photonic crystal nanocavities. *Optics Express* **15**, 17231 (2007).

262. Y. Akahane, T. Asano, B.-S. Song, and S. Noda. Fine-tuned high-Q photonic-crystal nanocavity. *Optics Express* **13**, 1202 (2005).

263. T. Tanabe, M. Notomi, E. Kuramochi, A. Shinya, and H. Taniyama. Trapping and delaying photons for one nanosecond in an ultrasmall high-Q photonic-crystal nanocavity. *Nature Photonics* **1**, 49 (2007).

264. E. Kuramochi, M. Notomi, S. Mitsugi, A. Shinya, T. Tanabe, and T. Watanabe. Ultrahigh-Q photonic crystal nanocavities realized by the local width modulation of a line defect. *Applied Physics Letters* **88**, 041112 (2006).

265. T. Yamamoto, M. Notomi, H. Taniyama, E. Kuramochi, Y. Yoshikawa, Y. Torii, and T. Kuga. Design of a high-Q air-slot cavity based on a width-modulated line-defect in a photonic crystal slab. *Optics Express* **16**, 13809 (2008).

266. Y. Takahashi, Y. Tanaka, H. Hagino, T. Sugiya, Y. Sato, T. Asano, and S. Noda. Design and demonstration of high-Q photonic heterostructure nanocavities suitable for integration. *Optics Express* **17**, 18093 (2009).

267. M. Barth, N. Nüsse, J. Stingl, B. Löchel, and O. Benson. Emission properties of high-Q silicon nitride photonic crystal heterostructure cavities. *Applied Physics Letters* **93**, 021112 (2008).

268. Y. Takahashi, H. Hagino, Y. Tanaka, B.-S. Song, T. Asano, and S. Noda. High-Q nanocavity with a 2-ns photon lifetime. *Optics Express* **15**, 17206 (2007).

269. S.-H. Kwon, T. Sünner, M. Kamp, and A. Forchel. Ultrahigh-Q photonic crystal cavity created by modulating air hole radius of a waveguide. *Optics Express* **16**, 4605 (2008).

270. M. Notomi and H. Taniyama. On-demand ultrahigh-Q cavity formation and photon pinning via dynamic waveguide tuning. *Optics Express* **16**, 18657 (2008).

271. M. Barth, N. Nüsse, B. Löchel, and O. Benson. Controlled coupling of a single-diamond nanocrystal to a photonic crystal cavity. *Optics Letters* **34**, 1108 (2009).

272. R. Sapienza, T. Coenen, J. Renger, M. Kuttge, N. F. van Hulst, and A. Polman. Deep-subwavelength imaging of the modal dispersion of light. *Nature Materials* **11**, 781 (2012).

273. N. Le Thomas, D. T. L. Alexander, M. Cantoni, W. Sigle, R. Houdré, and C. Hébert. Imaging of high-Q cavity optical modes by electron energy-loss microscopy. *Physical Review B* **87**, 155314 (2013).

274. N.-V.-Q. Tran, S. Combrié, and A. De Rossi. Directive emission from high-Q photonic crystal cavities through band folding. *Physical Review B* **79**, 041101 (2009).

275. S. Haddadi, L. Le-Gratiet, I. Sagnes, F. Raineri, A. Bazin, K. Bencheikh, J. A. Levenson, and A. M. Yacomotti. High quality beaming and efficient free-space coupling in L3 photonic crystal active nanocavities. *Optics Express* **20**, 18876 (2012).

276. M. Galli, S. L. Portalupi, M. Belotti, L. C. Andreani, L. O'Faolain, and T. F. Krauss. Light scattering and Fano resonances in high-Q photonic crystal nanocavities. *Applied Physics Letters* **94**, 071101 (2009).

277. U. Fano. Effects of configuration interaction on intensities and phase shifts. *Physical Review* **124**, 1866 (1961).

278. K. Hennessy, A. Badolato, A. Tamboli, P. M. Petroff, E. Hu, M. Atatüre, J. Dreiser, and A. Imamoglu. Tuning photonic crystal nanocavity modes by wet chemical digital etching. *Applied Physics Letters* **87**, 021108 (2005).

279. M. Barth, S. Schietinger, S. Fischer, J. Becker, N. Nüsse, T. Aichele, B. Löchel, C. Sönnichsen, and O. Benson. Nanoassembled plasmonic-photonic hybrid cavity for tailored light-matter coupling. *Nano Letters* **10**, 891 (2010).

280. J. Riedrich-Möller, L. Kipfstuhl, C. Hepp, E. Neu, C. Pauly, F. Mücklich, A. Baur, M. Wandt, S. Wolff, M. Fischer, S. Gsell, M. Schreck, and C. Becher. One- and two-dimensional photonic crystal micro-cavities in single crystal diamond. *Nature Nanotechnology* **7**, 69 (2012).

281. H. S. Lee, S. Kiravittaya, S. Kumar, J. D. Plumhof, L. Balet, L. H. Li, M. Francardi, A. Gerardino, A. Fiore, A. Rastelli, and O. G. Schmidt.

Local tuning of photonic crystal nanocavity modes by laser-assisted oxidation. *Applied Physics Letters* **95**, 191109 (2009).

282. E. Waks and J. Vučković. Coupled mode theory for photonic crystal cavity-waveguide interaction. *Optics Express* **13**, 5064 (2005).

283. Q. Li, T. Wang, Y. Su, M. Yan, and M. Qiu. Coupled mode theory analysis of mode-splitting in coupled cavity system. *Optics Express* **18**, 8367 (2010).

284. A. Faraon, E. Waks, D. Englund, I. Fushman, and J. Vučković. Efficient photonic crystal cavity-waveguide couplers. *Applied Physics Letters* **90**, 073102 (2007).

285. K. Mnaymneh, S. Frédérick, D. Dalacu, J. Lapointe, P. J. Poole, and R. L. Williams. Enhanced photonic crystal cavity-waveguide coupling using local slow-light engineering. *Optics Letters* **37**, 280 (2012).

286. O. Benson and F. Henneberger. *Semiconductor quantum bits*. Pan Stanford Publishing (2009).

287. A. N. Pikhtin, V. T. Prokopenko, V. S. Rondarev, and A. D. Yas'kov. Refraction of light in gallium phosphide. *Journal of Applied Spectroscopy* **27**, 1047 (1977).

288. G. A. Slack and S. F. Bartram. Thermal expansion of some diamondlike crystals. *Journal of Applied Physics* **46**, 89 (1975).

289. M. Bertolotti, V. Bogdanov, A. Ferrari, A. Jascow, N. Nazorova, A. Pikhtin, and L. Schirone. Temperature dependence of the refractive index in semiconductors. *Journal of the Optical Society of America B* **7**, 918 (1990).

290. G. Cocorullo, F. G. Della Corte, L. Moretti, I. Rendina, and A. Rubino. Measurement of the thermo-optic coefficient of a-Si:H at the wavelength of 1500 nm from room temperature to 200 °C. *Journal of Non-Crystalline Solids* **299**, 310 (2002).

291. P. Dean and D. Thomas. Intrinsic absorption-edge spectrum of gallium phosphide. *Physical Review* **150**, 690 (1966).

292. V. K. Subashiev and G. A. Chalikyan. The absorption spectrum of gallium phosphide between 2 and 3 eV. *Physica Status Solidi (B)* **13**, K91 (1966).

293. M. Belotti, J. F. Galisteo Lòpez, S. De Angelis, M. Galli, I. Maksymov, L. C. Andreani, D. Peyrade, and Y. Chen. All-optical switching in 2D silicon photonic crystals with low loss waveguides and optical cavities. *Optics Express* **16**, 11624 (2008).

294. X. Hu, P. Jiang, C. Ding, H. Yang, and Q. Gong. Picosecond and low-power all-optical switching based on an organic photonic-bandgap microcavity. *Nature Photonics* **2**, 185 (2008).

295. K. Nozaki, T. Tanabe, A. Shinya, S. Matsuo, T. Sato, H. Taniyama, and M. Notomi. Sub-femtojoule all-optical switching using a photonic-crystal nanocavity. *Nature Photonics* **4**, 477 (2010).

296. T. Tanabe, M. Notomi, S. Mitsugi, A. Shinya, and E. Kuramochi. All-optical switches on a silicon chip realized using photonic crystal nanocavities. *Applied Physics Letters* **87**, 151112 (2005).

297. Leo Merrill. Behavior of the AB2-type compounds at high pressures and high temperatures. *Journal of Physical and Chemical Reference Data* **11**, 1005 (1982).

298. M. Tmar, A. Gabriel, C. Chatillon, and I. Ansara. Critical analysis and optimization of the thermodynamic properties and phase diagrams in the III-V compounds: the In-P and Ga-P systems. *Journal of Crystal Growth* **68**, 557 (1984).

299. R. Weil. The elastic constants of gallium phosphide. *Journal of Applied Physics* **39**, 4049 (1968).

300. J. Crank, P. Nicolson, and D. R. Hartree. A practical method for numerical evaluation of solutions of partial differential equations of the heat-conduction type. *Mathematical Proceedings of the Cambridge Philosophical Society* **43**, 50 (2008).

301. M. R. Hestenes and E. Stiefel. Methods of conjugate gradients for solving linear systems. *Journal of Research of the National Bureau of Standards* **49**, 409 (1952).

302. D. Englund, A. Faraon, I. Fushman, N. Stoltz, P. Petroff, and J. Vučković. Controlling cavity reflectivity with a single quantum dot. *Nature* **450**, 857 (2007).

303. A. Faraon, A. Majumdar, D. Englund, E. Kim, M. Bajcsy, and J. Vučković. Integrated quantum optical networks based on quantum dots and photonic crystals. *New Journal of Physics* **13**, 055025 (2011).

304. E. Moreau, I. Robert, J. M. Gerard, I. Abram, L. Manin, and V. Thierry-Mieg. Single-mode solid-state single photon source based on isolated quantum dots in pillar microcavities. *Applied Physics Letters* **79**, 2865 (2001).

305. G. Solomon, M. Pelton, and Y. Yamamoto. Single-mode spontaneous emission from a single quantum dot in a Three-Dimensional Microcavity. *Physical Review Letters* **86**, 3903 (2001).

306. D. Englund, D. Fattal, E. Waks, G. Solomon, B. Zhang, T. Nakaoka, Y. Arakawa, Y. Yamamoto, and J. Vučković. Controlling the spontaneous emission rate of single quantum Dots in a two-dimensional photonic crystal. *Physical Review Letters* **95**, 013904 (2005).

307. D. G. Gevaux, A. J. Bennett, R. M. Stevenson, A. J. Shields, P. Atkinson, J. Griffiths, D. Anderson, G. A. C. Jones, and D. A. Ritchie. Enhancement and suppression of spontaneous emission by temperature tuning InAs quantum dots to photonic crystal cavities. *Applied Physics Letters* **88**, 131101 (2006).

308. E. Stock, F. Albert, C. Hopfmann, M. Lermer, C. Schneider, S. Höfling, A. Forchel, M. Kamp, and S. Reitzenstein. On-chip quantum optics with quantum dot microcavities. *Advanced materials (Deerfield Beach, Fla.)* **25**, 707 (2013).

309. C. Santori, D. Fattal, J. Vučković, G. S. Solomon, and Y. Yamamoto. Indistinguishable photons from a single-photon device. *Nature* **419**, 594 (2002).

310. T. Legero, T. Wilk, A. Kuhn, and G. Rempe. Time-resolved two-photon quantum interference. *Applied Physics B: Lasers and Optics* **77**, 797 (2003).

311. C. K. Hong, Z. Y. Ou, and L. Mandel. Measurement of subpicosecond time intervals between two photons by interference. *Physical Review Letters* **59**, 2044 (1987).

312. J. Beugnon, M. P. A. Jones, J. Dingjan, B. Darquié, G. Messin, A. Browaeys, and P. Grangier. Quantum interference between two single photons emitted by independently trapped atoms. *Nature* **440**, 779 (2006).

313. S. V. Polyakov, A. Muller, E. B. Flagg, A. Ling, N. Borjemscaia, E. Van Keuren, A. Migdall, and G. S. Solomon. Coalescence of single photons emitted by disparate single-photon sources: the example of InAs Quantum Dots and parametric down-conversion sources. *Physical Review Letters* **107**, 157402 (2011).

314. M. Gündoğan, P. M. Ledingham, A. Almasi, M. Cristiani, and H. de Riedmatten. Quantum storage of a photonic polarization qubit in a solid. *Physical Review Letters* **108**, 190504 (2012).

315. L. Slodička, G. Hétet, N. Röck, P. Schindler, M. Hennrich, and R. Blatt. Atom-atom entanglement by single-photon detection. *Physical Review Letters* **110**, 083603 (2013).

316. V. Baumann, F. Stumpf, T. Steinl, A. Forchel, C. Schneider, S. Höfling, and M. Kamp. Site-controlled growth of InP/GaInP quantum dots on GaAs substrates. *Nanotechnology* **23**, 375301 (2012).

317. A. Faraon, P. E. Barclay, C. Santori, K.-M. C. Fu, and R. G. Beausoleil. Resonant enhancement of the zero-phonon emission from a colour centre in a diamond cavity. *Nature Photonics* **5**, 301 (2011).

318. J. Riedrich-Möller, L. Kipfstuhl, C. Hepp, E. Neu, C. Pauly, F. Mücklich, A. Baur, M. Wandt, S. Wolff, M. Fischer, S. Gsell, M. Schreck, and C. Becher. One- and two-dimensional photonic crystal microcavities in single crystal diamond. *Nature Nanotechnology* **7**, 69 (2012).

319. A. Faraon, C. Santori, Z. Huang, V. M. Acosta, and R. G. Beausoleil. Coupling of nitrogen-vacancy centers to photonic crystal cavities in monocrystalline diamond. *Physical Review Letters* **109**, 033604 (2012).

320. J. C. Lee, I. Aharonovich, A. P. Magyar, F. Rol, and E. L. Hu. Coupling of silicon-vacancy centers to a single crystal diamond cavity. *Optics Express* **20**, 8891 (2012).

321. B. J. M. Hausmann, B. Shields, Q. Quan, P. Maletinsky, M. McCutcheon, J. T. Choy, T. M. Babinec, A. Kubanek, A. Yacoby, M. D. Lukin, and M. Loncar. Integrated diamond networks for quantum nanophotonics. *Nano Letters* **12**, 1578 (2012).

322. B. K. Ofori-Okai, S. Pezzagna, K. Chang, M. Loretz, R. Schirhagl, Y. Tao, B. A. Moores, K. Groot-Berning, J. Meijer, and C. L. Degen. Spin properties of very shallow nitrogen vacancy defects in diamond. *Physical Review B* **86**, 081406 (2012).

323. C. Roos. Moving traps offer fast delivery of cold ions. *Physics* **5**, 94 (2012).

324. A. Walther, F. Ziesel, T. Ruster, S. T. Dawkins, K. Ott, M. Hettrich, K. Singer, F. Schmidt-Kaler, and U. Poschinger. Controlling fast transport of cold trapped ions. *Physical Review Letters* **109**, 080501 (2012).

325. P. Spinicelli, A. Dréau, L. Rondin, F. Silva, J. Achard, S. Xavier, S. Bansropun, T. Debuisschert, S. Pezzagna, J. Meijer, V. Jacques, and J.-F. Roch. Engineered arrays of nitrogen-vacancy color centers in diamond based on implantation of CN⁻ molecules through nanoapertures. *New Journal of Physics* **13**, 025014 (2011).

326. N. Spagnolo, L. Aparo, C. Vitelli, A. Crespi, R. Ramponi, R. Osellame, P. Mataloni, and F. Sciarrino. Quantum interferometry with three-dimensional geometry. *Scientific Reports* **2**, 862 (2012).

327. T. van der Sar, E. C. Heeres, G. M. Dmochowski, G. de Lange, L. Robledo, T. H. Oosterkamp, and R. Hanson. Nanopositioning of a diamond nanocrystal containing a single nitrogen-vacancy defect center. *Applied Physics Letters* **94**, 173104 (2009).

328. E. Ampem-Lassen, D. A. Simpson, B. C. Gibson, S. Trpkovski, F. M. Hossain, S. T. Huntington, K. Ganesan, L. C. Hollenberg, and S. Prawer. Nano-manipulation of diamond-based single photon sources. *Optics Express* **17**, 11287 (2009).

329. A. Huck, S. Kumar, A. Shakoor, and U. L. Andersen. Controlled coupling of a single nitrogen-vacancy center to a silver nanowire. *Physical Review Letters* **106**, 096801 (2011).

330. A. W. Schell, G. Kewes, T. Schröder, J. Wolters, T. Aichele, and O. Benson. A scanning probe-based pick-and-place procedure for assembly of integrated quantum optical hybrid devices. *The Review of Scientific Instruments* **82**, 073709 (2011).

331. S. Schietinger, T. Schröder, and O. Benson. One-by-one coupling of single defect centers in nanodiamonds to high-Q modes of an optical microresonator. *Nano Letters* **8**, 3911 (2008).

332. M. Gregor, R. Henze, T. Schröder, and O. Benson. On-demand positioning of a preselected quantum emitter on a fiber-coupled toroidal microresonator. *Applied Physics Letters* **95**, 153110 (2009).

333. S. Schietinger, M. Barth, T. Aichele, and O. Benson. Plasmon-enhanced single photon emission from a nanoassembled metal-diamond hybrid structure at room temperature. *Nano Letters* **9**, 1694 (2009).

334. A. W. Schell, G. Kewes, T. Hanke, A. Leitenstorfer, R. Bratschitsch, O. Benson, and T. Aichele. Single defect centers in diamond nanocrystals as quantum probes for plasmonic nanostructures. *Optics Express* **19**, 7914 (2011).

335. T. Schröder, A. W. Schell, G. Kewes, T. Aichele, and O. Benson. Fiber-integrated diamond-based single photon source. *Nano Letters* **11**, 198 (2011).

336. K. Hennessy, A. Badolato, M. Winger, D. Gerace, M. Atatüre, S. Gulde, S. Fält, E. L. Hu, and A. Imamoglu. Quantum nature of a strongly coupled single quantum dot-cavity system. *Nature* **44**, 896 (2007).

337. C. Santori, P. E. Barclay, K.-M. C. Fu, R. G. Beausoleil, S. Spillane, and M. Fisch. Nanophotonics for quantum optics using nitrogen-vacancy centers in diamond. *Nanotechnology*, **21**(27), 274008 (2010).

338. C.-H. Su, A. D. Greentree, and L. C. L. Hollenberg. Towards a picosecond transform-limited nitrogen-vacancy based single photon source. *Optics Express* **16**, 6247 (2008).

339. P. E. Barclay, C. Santori, K.-M. Fu, R. G. Beausoleil, and O. Painter. Coherent interference effects in a nano-assembled diamond NV center cavity-QED system. *Optics Express* **17**, 8081 (2009).

340. C.-H. Su, A. Greentree, W. Munro, K. Nemoto, and L. Hollenberg. High-speed quantum gates with cavity quantum electrodynamics. *Physical Review A* **78**, 062336 (2008).

341. D. Englund, A. Faraon, B. Zhang, Y. Yamamoto, and J. Vučković. Generation and transfer of single photons on a photonic crystal chip. *Optics Express* **15**, 5550 (2007).

342. A. Faraon, I. Fushman, D. Englund, N. Stoltz, P. Petroff, and J. Vučković. Dipole induced transparency in waveguide coupled photonic crystal cavities. *Optics Express* **16**, 12154 (2008).

343. A. Young, C. Y. Hu, L. Marseglia, J. P. Harrison, J. L. O'Brien, and J. G. Rarity. Cavity enhanced spin measurement of the ground state spin of an NV center in diamond. *New Journal of Physics* **11**, 013007 (2009).

344. H. Mabuchi and A. C. Doherty. Cavity quantum electrodynamics: coherence in context. *Science* **298**, 1372 (2002).

345. J. Cirac, P. Zoller, H. Kimble, and H. Mabuchi. Quantum state transfer and entanglement distribution among distant nodes in a quantum network. *Physical Review Letters* **78**, 3221 (1997).

346. L.-M. Duan and H. Kimble. Scalable photonic quantum computation through cavity-assisted interactions. *Physical Review Letters* **92**, 127902 (2004).

347. S. Ritter, C. Nölleke, C. Hahn, A. Reiserer, A. Neuzner, M. Uphoff, M. Mücke, E. Figueroa, J. Bochmann, and G. Rempe. An elementary quantum network of single atoms in optical cavities. *Nature* **484**, 195 (2012).

348. T. Volz, A. Reinhard, M. Winger, A. Badolato, K. J. Hennessy, E. L. Hu, and A. Imamoglu. Ultrafast all-optical switching by single photons. *Nature Photonics* **6**, 607 (2012).

349. T. Yoshie, A. Scherer, J. Hendrickson, G. Khitrova, H. M. Gibbs, G. Rupper, C. Ell, O. B. Shchekin, and D. G. Deppe. Vacuum Rabi splitting with a single quantum dot in a photonic crystal nanocavity. *Nature* **432**, 200 (2004).

350. E. Peter, P. Senellart, D. Martrou, A. Lemaître, J. Hours, J. Gérard, and J. Bloch. Exciton-photon strong-coupling regime for a single Quantum Dot embedded in a microcavity. *Physical Review Letters* **95**, 067401 (2005).

351. J. P. Reithmaier, G. Sek, A. Löffler, C. Hofmann, S. Kuhn, S. Reitzenstein, L. V. Keldysh, V. D. Kulakovskii, T. L. Reinecke, and A. Forchel. Strong coupling in a single quantum dot-semiconductor microcavity system. *Nature* **432**, 197 (2004).

352. J. Christen and D. Bimberg. Line shapes of intersubband and excitonic recombination in quantum wells: influence of final-state interaction,

statistical broadening, and momentum conservation. *Physical Review B* **42**, 7213 (1990).

353. M. Bajcsy, A. Majumdar, A. Rundquist, and J. Vučković. Photon blockade with a four-level quantum emitter coupled to a photonic-crystal nanocavity. *New Journal of Physics* **15**, 025014 (2013).

354. A. Majumdar, M. Bajcsy, and J. Vučković. Probing the ladder of dressed states and nonclassical light generation in quantum-dot cavity QED. *Physical Review A* **85**, 041801 (2012).

355. A. Majumdar, P. Kaer, M. Bajcsy, E. D. Kim, K. G. Lagoudakis, A. Rundquist, and J. Vučković. Proposed coupling of an electron spin in a semiconductor Quantum Dot to a nanosize optical cavity. *Physical Review Letters* **111**, 027402 (2013).

356. W. L. Yang, Z. Q. Yin, Z. Y. Xu, M. Feng, and C. H. Oh. Quantum dynamics and quantum state transfer between separated nitrogen-vacancy centers embedded in photonic crystal cavities. *Physical Review A* **84**, 043849 (2011).

357. W. L. Yang, Z. Q. Yin, Z. Y. Xu, M. Feng, and J. F. Du. One-step implementation of multiqubit conditional phase gating with nitrogen-vacancy centers coupled to a high-Q silica microsphere cavity. *Applied Physics Letters* **96**, 241113 (2010).

358. W. Yang, Z. Xu, M. Feng, and J. Du. Entanglement of separate nitrogen-vacancy centers coupled to a whispering-gallery mode cavity. *New Journal of Physics* **12**, 113039 (2010).

359. S. Liu, J. Li, R. Yu, and Y. Wu. Achieving maximum entanglement between two nitrogen-vacancy centers coupling to a whispering-gallery-mode microresonator. *Optics Express* **21**, 3501 (2013).

360. M. W. McCutcheon and M. Loncar. Design of a silicon nitride photonic crystal nanocavity with a quality factor of one million for coupling to a diamond nanocrystal. *Optics Express* **16**, 19136 (2008).

361. A. Einstein, B. Podolsky, and N. Rosen. Can quantum-mechanical description of physical reality be considered complete? *Physical Review* **47**, 777 (1935).

362. J. S. Bell. On the Einstein Podolsky Rosen paradox. *Physics* **1**, 175 (1964).

363. J. Clauser, M. Horne, A. Shimony, and R. Holt. Proposed experiment to test local hidden-variable theories. *Physical Review Letters* **23**, 880 (1969).

364. A. Aspect, P. Grangier, and G. Roger. Experimental tests of realistic local theories via Bell's theorem. *Physical Review Letters* **47**, 460 (1981).

365. A. Aspect, P. Grangier, and G. Roger. Experimental realization of Einstein-Podolsky-Rosen-Bohm Gedanken experiment: a new violation of Bell's inequalities. *Physical Review Letters* **49**, 91 (1982).

366. R. P. Feynman. Simulating physics with computers. *International Journal of Theoretical Physics* **21**, 467 (1982).

367. H. Kimble, Y. Levin, A. Matsko, K. Thorne, and S. Vyatchanin. Conversion of conventional gravitational-wave interferometers into quantum nondemolition interferometers by modifying their input and/or output optics. *Physical Review D* **65**, 022002 (2001).

368. T. Wilk, A. Gaëtan, C. Evellin, J. Wolters, Y. Miroshnychenko, P. Grangier, and A. Browaeys. Entanglement of two individual neutral atoms using rydberg blockade. *Physical Review Letters* **104**, 010502 (2010).

369. A. J. Berkley, H. Xu, R. C. Ramos, M. A. Gubrud, F. W. Strauch, P. R. Johnson, J. R. Anderson, A. J. Dragt, C. J. Lobb, and F. C. Wellstood. Entangled macroscopic quantum states in two superconducting qubits. *Science* **300**, 1548 (2003).

370. J. Hofmann, M. Krug, N. Ortegel, L. Gérard, M. Weber, W. Rosenfeld, and H. Weinfurter. Heralded entanglement between widely separated atoms. *Science* **337**, 72 (2012).

371. R. Blatt and C. F. Roos. Quantum simulations with trapped ions. *Nature Physics* **8**, 277 (2012).

372. A. Stute, B. Casabone, B. Brandstätter, K. Friebe, T. E. Northup, and R. Blatt. Quantum-state transfer from an ion to a photon. *Nature Photonics* **7**, 219 (2013).

373. G. D. Fuchs, G. Burkard, P. V. Klimov, and D. D. Awschalom. A quantum memory intrinsic to single nitrogen-vacancy centres in diamond. *Nature Physics* **7**, 789 (2011).

374. K. De Greve, L. Yu, P. L. McMahon, J. S. Pelc, C. M. Natarajan, N. Y. Kim, E. Abe, S. Maier, C. Schneider, M. Kamp, S. Höfling, R. H. Hadfield, A. Forchel, M. M. Fejer, and Y. Yamamoto. Quantum-dot spin-photon entanglement via frequency downconversion to telecom wavelength. *Nature* **491**, 421 (2012).

375. H. J. Kimble. The quantum internet. *Nature* **453**, 1023 (2008).

376. S. Barrett and P. Kok. Efficient high-fidelity quantum computation using matter qubits and linear optics. *Physical Review A* **71**, 060310 (2005).

377. J. Wolters, J. Kabuß, A. Knorr, and O. Benson. Deterministic and robust entanglement of nitrogen-vacancy centers using low-Q photonic-crystal cavities. *Physical Review A* **89**, 060303 (2014).

378. J. P. Sprengers, A. Gaggero, D. Sahin, S. Jahanmirinejad, G. Frucci, F. Mattioli, R. Leoni, J. Beetz, M. Lermer, M. Kamp, S. Höfling, R. Sanjines, and A. Fiore. Waveguide superconducting single-photon detectors for integrated quantum photonic circuits. *Applied Physics Letters* **99**, 181110 (2011).

379. A. Imamoglu, D. D. Awschalom, G. Burkard, D. P. DiVincenzo, D. Loss, M. Sherwin, and A. Small. Quantum information processing using quantum dot spins and cavity QED. *Physical Review Letters* **83**, 4204 (1999).

380. L. Robledo, H. Bernien, I. van Weperen, and R. Hanson. Control and coherence of the optical transition of single nitrogen vacancy centers in diamond. *Physical Review Letters* **105**, 177403 (2010).

381. D. F. V. James, P. G. Kwiat, W. J. Munro, and A. G. White. Measurement of qubits. *Physical Review A* **64**, 052312 (2001).

382. M. A. Nielsen and I. L. Chuang. *Quantum Computation and Quantum Information*. Cambridge University Press (2010).

383. C. Barth, J. Wolters, A. W. Schell, J. Probst, M. Schoengen, B. Löchel, S. Kowarik, and O. Benson. Miniaturized Bragg-grating couplers for SiN-photonic crystal slabs. *Optics Express* **23**, 9803 (2015).

384. T. Heindel. Elektrisch gepumpte Quantenpunkt-Einzelphotonenquellen für die Quantenkommunikation, PhD Thesis, Universität Würzburg, Fakultät für Physik und Astronomie, 2013.

List of Abbreviations

AC	anthracene
AFM	atomic force microscope
AIP	American Institute of Physics Publishing LLC
AOM	acousto-optic modulator
APD	avalanche photo diode
APS	American Physical Society Publishing
BS	non-polarizing beam splitter
CCD	charge coupled device
CMOS	complementary metal oxide semiconductor
CPT	coherent population trapping
CVD	chemical vapor deposition
cw	continuous wave
DBATT	dibenzanthanthrene
DBT	dibenzoterrylene
DLW	direct laser writing
DM	dichroic mirror
EIT	electromagnetic induced transparency
ENDOR	electro nuclear double resonance
EO	entanglement operation
EPR	Einstein-Podolsky-Rosen
FDTD	finite-difference time-domain method
FWHM	full width half maximum
HBT	Hanbury Brown and Twiss setup
HTHP	high pressure and high temperature
ISC	inter system crossing
IOP	Institute of Physics Publishing
LED	light emitting diode
LP	long pass filter
LPCVD	low pressure chemical vapor deposition

MIT	massachusetts institute of technology
MPB	MIT photonic bands package
MW	microwave
NA	numerical aperture
NMR	nuclear magnetic resonance
NV	nitrogen vacancy
ODMR	optically detected magnetic resonance
OSA	The Optical Society Publishing
PC	photonic crystal
PL	photoluminescence
PMMA	polymethylmethacrylate
PSB	phonon side-band
PVA	polyvinyl alcohol
QD	quantum dot
QED	quantum electrodynamics
QKD	quantum key distribution
RWA	rotating wave approximation
SE	spontaneous emission
SEM	scanning electron microscop
SF	spatial filter
SIL	solid immersion lens
SiV	silicon-vacancy center
SNR	signal-to-noise ratio
SRT	stimulated Raman transition
STIRAP	stimulated Raman adiabatic passage
TE	transverse electric field
TM	transverse magnetic field
TTTR	time-tagged time-resolved
VLSI	very large scale integration
WGM	whispering gallery modes
X	exciton
XX	biexciton
ZPL	zero phonon line

Index